Machine Learning for Industrial Applications

Scrivener Publishing
100 Cummings Center, Suite 541J
Beverly, MA 01915-6106

Next-Generation Computing and Communication Engineering

Series Editors: Dr. G. R. Kanagachidambaresan and Dr. Kolla Bhanu Prakash

Edge computing has become an active research field supporting low processing power, real-time response time, and more resource capacity than IoT and mobile devices. It has also been considered to effectively mitigate loads on data centers, to assist artificial intelligence (AI) services, and to increase 5G services. Edge computing applications along with the IoT field are essential technical directions in order to open the door to new opportunities enabling smart homes, smart hospitals, smart cities, smart vehicles, smart wearables, smart supply chain, e-health, automation, and a variety of other smart environments.

However, any developments are made more challenging because the involvement of multi-domain technology creates new problems for researchers. Therefore, in order to help meet the challenge, this book series concentrates on next generation computing and communication methodologies involving smart and ambient environment design. It is an effective publishing platform for monographs, handbooks, and edited volumes on Industry 4.0, agriculture, smart city development, new computing and communication paradigms. Although the series mainly focuses on design, it also addresses analytics and investigation of industry-related real-time problems.

Publishers at Scrivener
Martin Scrivener (martin@scrivenerpublishing.com)
Phillip Carmical (pcarmical@scrivenerpublishing.com)

Machine Learning for Industrial Applications

Kolla Bhanu Prakash
Department of Computer Science & Engineering, K. L. University, Vaddeswaram, India

Scrivener Publishing

WILEY

This edition first published 2024 by John Wiley & Sons, Inc., 111 River Street, Hoboken, NJ 07030, USA and Scrivener Publishing LLC, 100 Cummings Center, Suite 541J, Beverly, MA 01915, USA

For more information about Scrivener publications please visit www.scrivenerpublishing.com.

Wiley Global Headquarters
111 River Street, Hoboken, NJ 07030, USA

For details of our global editorial offices, customer services, and more information about Wiley products visit us at www.wiley.com.

Library of Congress Cataloging-in-Publication Data

ISBN 978-1-394-26896-2

Cover image: Pixabay.Com
Cover design by Russell Richardson

Set in size of 11pt and Minion Pro by Manila Typesetting Company, Makati, Philippines

Printed in the USA

10 9 8 7 6 5 4 3 2 1

Dedicated to my parents, Sri. Kolla Narayana Rao
& Smt. Kolla Uma Maheswari
and my wife, Mrs. M. V. Prasanna Lakshmi

Contents

Preface

Welcome to the exciting world of machine learning! In recent years, machine learning has rapidly transformed from a niche field within computer science to a fundamental technology shaping various aspects of our lives. Whether you realize it or not, machine learning algorithms are at work behind the scenes, powering recommendation systems, autonomous vehicles, virtual assistants, medical diagnostics, and much more. This book is designed to serve as your comprehensive guide to understanding the principles, algorithms, and applications of machine learning. Whether a student diving into this field for the first time, a seasoned professional looking to broaden your skillset, or an enthusiast eager to explore cutting-edge advancements, this book has something for you. In this preface, we'll outline the objectives of this book, discuss its structure, and offer some guidance on how to get the most out of your journey through its pages.

Our primary goal is to demystify machine learning and make it accessible to a wide audience. We aim to provide a solid foundation in the fundamental concepts of machine learning, covering both the theoretical underpinnings and practical applications. Whether you're interested in supervised learning, unsupervised learning, reinforcement learning, or innovative techniques like deep learning, you'll find comprehensive coverage here. Throughout the book, we emphasize a hands-on approach. We firmly believe that the best way to learn machine learning is by doing. Therefore, we include numerous examples, exercises, and real-world case studies to reinforce your understanding and practical skills.

I am deeply grateful for the dedicated support and valuable assistance rendered by Martin Scrivener and the Scrivener Publishing team during this book.

Dr. Kolla Bhanu Prakash
June 2024

1

Overview of Machine Learning

1.1 Introduction

Machine learning is a branch of artificial intelligence (AI) that focuses on developing algorithms and models that enable computers to learn and make predictions or decisions without being explicitly programmed. It is a field that has gained significant attention and prominence due to its ability to solve complex problems and make sense of large amounts of data.

Machine learning teaches computers how to learn from data and improve their performance over time. This is achieved through various techniques and algorithms that allow machines to recognize patterns, make predictions, and adapt their behavior based on their input.

There are several key concepts and components within machine learning:

Data: Data are the foundation of machine learning. Algorithms learn from data, which can be in the form of text, images, videos, numerical values, or any other structured or unstructured format.
Features: Features are the attributes or characteristics of the data that are used by algorithms to make predictions or classifications. Selecting relevant features is crucial for the success of a machine learning model.
Labels: In supervised learning, which is a common type of machine learning, data are labeled with the correct output. The algorithm learns from the labeled data to predict new, unseen data.
Algorithms: Machine learning algorithms are mathematical models that process and learn from data. These algorithms can be categorized into different types, such as regression, classification, clustering, and reinforcement learning, each designed for specific tasks.
Training: Training a machine learning model involves feeding it labeled data and allowing it to adjust its internal parameters to minimize the difference between its predictions and the actual labels.

Kolla Bhanu Prakash. Machine Learning for Industrial Applications, (1–20) © 2024 Scrivener Publishing LLC

Testing and Validation: The model's performance is evaluated on new, unseen data to assess its generalization ability after training. This helps ensure the model is balanced (performing well on training data but poorly on new data).
Supervised, Unsupervised, and Reinforcement Learning: These are the main categories of machine learning. In supervised learning, the algorithm learns from labeled data to make predictions. In unsupervised learning, the algorithm identifies patterns and relationships in data without explicit labels. Reinforcement learning involves training agents to take action in an environment to maximize a reward.
Overfitting and Underfitting: Overfitting occurs when a model learns the training data too well and performs poorly on new data. Conversely, underfitting happens when a model is too simplistic to capture the underlying patterns in the data.
Hyperparameters: These are parameters that are set before training a model, such as the learning rate or the number of hidden layers in a neural network. Tuning hyperparameters is an essential aspect of optimizing model performance.
Deep Learning: Deep learning is a subset of machine learning focusing on multiple layers of neural networks. It has revolutionized fields like computer vision, natural language processing, and speech recognition.

Machine learning has many applications, including image and speech recognition, medical diagnosis, recommendation systems, fraud detection, and autonomous vehicles. As more data become available and computing power increases, machine learning advances and shapes various aspects of our lives.

Machine learning is a subset of AI that focuses on developing algorithms and statistical models, allowing computers to learn and improve their performance on specific tasks without being explicitly programmed. The core idea behind machine learning is to enable systems to recognize patterns, make predictions, and learn from data, mimicking how humans learn from experience.

In traditional programming, a set of rules and instructions is provided to a computer to solve a problem or perform a task. In contrast, machine learning algorithms learn from data, adjusting their parameters and improving their performance based on the patterns they identify within the data. This process is often referred to as "training" the model.

Machine learning can be broadly categorized into three types:

Supervised Learning: In supervised learning, the algorithm is provided with labeled data, where each data point has an associated "correct" output. The model learns by comparing its predictions with the actual labels and adjusting its parameters accordingly. It is commonly used for tasks like classification and regression.
Unsupervised Learning: Unsupervised learning involves training the algorithm on unlabeled data, where there are no predefined outputs. The model learns to identify patterns and structures within the data without explicit guidance. Clustering and dimensionality reduction are typical applications of unsupervised learning.
Reinforcement Learning: Reinforcement learning is concerned with training agents to interact with an environment and learn from the feedback they receive. The agent's actions lead to rewards or penalties, helping it improve its decision-making process over time.

Machine learning has many applications across industries, including finance, healthcare, marketing, robotics, natural language processing, and image recognition. Its ability to extract valuable insights from vast amounts of data has significantly impacted industry development, enabling businesses to make data-driven decisions, optimize processes, enhance user experiences, and discover new opportunities for growth and innovation.

However, the successful implementation of machine learning requires careful data preparation, feature engineering, model selection, and ongoing monitoring to ensure its accuracy, reliability, and ethical use. As the field continues to evolve, advancements in deep learning, neural networks, and other techniques are further pushing the boundaries of what machine learning can achieve, making it a critical component in the advancement of technology and industry in the modern world.

Through getting to know an example from design inputs, the gadget concentrating on calculation predicts and performs errands completely based absolutely at the found example and not a predefined programming arrangement. Framework acquiring information is a way-of-life rescuer in a few cases where applying severe calculations is absurd. It will examine the spic-and-span technique from going before styles and executing the expertise.

One of the devices concentrating on applications that we are acquainted with is the way our email transporters help us address spam. Garbage mail channels utilize a calculation to become mindful of and pass approaching garbage messages to your spam organizer. Various web-based business bunches also use machine getting-to-know calculations related to various IT insurance devices to forestall misrepresentation and upgrade their recommendation motor performance [1].

1.2 Sorts of Machine Learning

Depending on your concern assertion, you can utilize both of the three methods to instruct your machine:

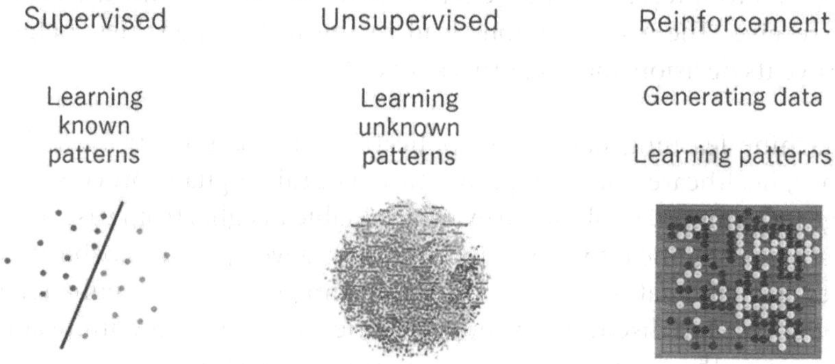

1.3 Regulated Gaining Knowledge of Dog and Human

Regulated device dominating should be carried out to datasets wherein the mark/class of every data is perceived. Allow us to reflect that we genuinely wish to train our system to distinguish between images of a human and a dog. Imagine that we have a collection of images that are categorized as either human or dog (human annotators carry out marking to ensure a greater top-notch data). Presently, we can utilize this dataset and data examples to prepare our calculation to concentrate on the appropriate way. When our arrangement of rules learns the method for grouping photos, we will utilize it on extraordinary datasets that are expecting the mark of any new realities factor.

1.4 Solo Learning

As you could wager from the call, the unaided device acquiring information is without any administering guidelines or names. We simply give our gadget a lot of realities and the attributes of every realities piece. As an example, thinking ahead of time occurrence, we basically took care of a few pictures (of people and pups) to our device giving each photograph a capacity. The qualities of individuals can be comparative and uncommon from young doggies. Utilizing these attributes, we will teach our gadget to bunch data into classes. An unmanaged variation of "type" is ordered "gathering." In bunching, we do not have any marks. The association of the datasets is dependent absolutely on ordinary characteristics.

1.5 Support Mastering

In support, realizing that there are no examples or attributes, there is essentially a stop point–skip or fall flat. To perceive this better, consider the occasion of acquiring information to play chess. After each game, the gadget is educated regarding the success/misfortune notoriety. In this sort of case, our gadget does now not have each stream named as "legitimate" or "erroneous," yet least complex has the final product. As our calculation plays more computer games eventually of the tutoring, it will safeguard giving bigger "loads" (importance) to the blend of those activities that prompted a win [2].

1.6 Bundles or Applications of Machine Learning

Contraption dominating is a popular expression for most recent age, and it is far developing startlingly each day. We are the use of gadget learning in our step by step life even without understanding it, which incorporates Google Maps, Google Assistant, Alexa, etc. The following are some of the greatest moving genuine worldwide projects of gadget learning.

1.6.1 Photograph Reputation

Photo notoriety is one of the most widely recognized bundles of gadget acquiring information. It is miles used to find objects, individuals, areas, virtual pics, etc. The famous use instance of photograph ubiquity and face recognition is programmed buddy labeling thought.

Facebook (FB) presents us a component of vehicle amigo labeling tips. At whatever point, we add a photograph with our facebook companions; then, at that point, we mechanically get a labeling motivation with a call; and the innovation toward the rear of that is machine concentrating on its face recognition and notoriety calculation.

It is far based on the FB challenge named "Profound Face" that is at risk for face acknowledgment and man or lady character in the photograph.

1.6.2 Discourse Recognition

Indeed, even as utilizing Google, we get a decision of "look for by utilizing voice," which comes underneath discourse acknowledgment, and it is a renowned utilization of machine dominating.

Discourse prominence is a process for changing over voice guidelines into message; furthermore it is insinuated as "Talk to message" or "PC talk acknowledgment." As of now, gadget learning calculations are generally utilized by different bundles of discourse notoriety. Google Assistant, Siri, Cortana, and Alexa are the utilization of discourse ubiquity period to conform to voice orders.

1.6.3 Traffic Prediction

To go to a spic-and-span region, we take the assistance of Google Maps, which shows us the right way with the briefest way and predicts the site guests' conditions.

It predicts the guests' conditions, which incorporate whether or not traffic is cleared, slow-moving, or firmly blocked with the assistance of ways:

- Genuine time district of the car from Google Maps application and sensors.
- Normal time has required on past days simultaneously.

Any individual who is utilizing Google Maps is helping this application to make it higher. It takes records from the customer and sends it lower back to its dataset to upgrade its exhibition.

1.6.4 Item Recommendations

Gadget dominating is widely used by different web-based business and diversion bunches, which incorporates Amazon, Netflix, etc., for item proposals to the buyer. On each event, we search for some item on Amazon; then, at that point, we began getting a promotion for the indistinguishable item simultaneously as web surfing at a similar program; and this is because of machine examining.

Google knows about the individual interest in the utilization of different contraption dominating calculations and demonstrates the item according to customer leisure activity.

As comparative, while we use Netflix, we find a couple of suggestions for amusement series, films, etc., and this is, furthermore, performed with the help of framework examining.

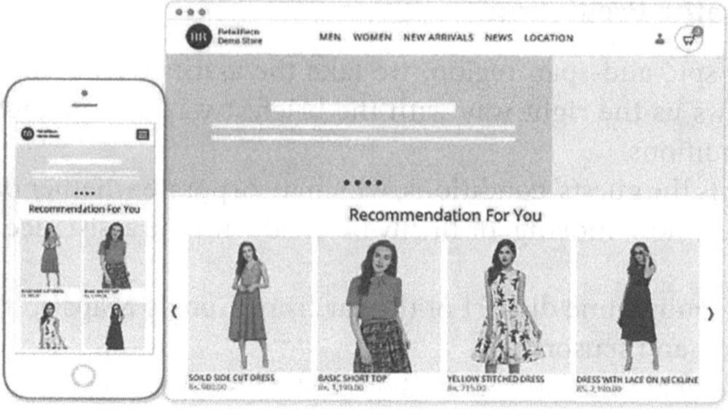

1.6.5 Self-Using Vehicles

One of the greatest energizing utilizations of machine considering is self-riding vehicles. Contraption acquiring information fills a gigantic role in self-riding vehicles. Tesla, the greatest famous auto creation association, is dealing with a self-utilizing vehicle. It is the utilization of an unmonitored dominating procedure to instruct the vehicle models to run over individuals and articles simultaneously as driving.

1.6.6 Electronic Mail Unsolicited Mail And Malware Filtering

On each event, we get hold of a pristine email, and it is miles sifted precisely as basic, regular, and garbage mail. We ordinarily get a crucial mail in our inbox with the significant image and spam messages in our garbage mail holder, and the time behind this is framework learning. The following are a couple of garbage mail channels utilized by Gmail:

- ➢ Content channel out
- ➢ Header sift through
- ➢ Favored boycotts sift through
- ➢ Rules based absolutely channels
- ➢ Authorization channels

A couple of machine dominating calculations including multi-layer perceptron, decision tree, and naïve Bayes classifier are utilized for electronic mail spontaneous mail sifting and malware location.

1.6.7 Computerized Private Assistant

We have different computerized non-public associates including Google Assistant, Alexa, Cortana, and Siri. As the call shows, they help us in tracking down the insights the utilization of our voice guidance. These associates can help us in different techniques simply through our voice orders comprising playing tune, calling somebody, beginning an email, Scheduling an arrangement, etc.

These menial helpers use gadget concentrating on calculations as a fundamental part.

Those partners record our voice directions, send them over to the server on a cloud, interpret it involving machine learning calculations, and represent that explanation.

1.6.8 Online Fraud Detection

Framework dominating is making our online exchanges secure and comfortable by distinguishing extortion exchanges. Each time we do a couple of online exchange, there might be different methodologies that a fake exchange can take area comprehensive of false bills, artificial ids, and taking cash inside the center of an exchange. So, you can stagger on this; feed ahead neural people group permits us with the guide of checking whether it is a real exchange or an extortion exchange.

For each real exchange, the result is changed into a couple of hash esteems, and those qualities become the contribution for the accompanying circular. For each real exchange, there is a chosen test that gets changed for the extortion exchange; therefore, it distinguishes it and makes our online exchanges additional comfortable.

1.6.9 Securities Exchange Buying and Selling

Contraption dominating is widely utilized in stock commercial center exchanging. Inside the stock commercial center, there might be continually a danger of ups and downs in shares, so for this machine, acquiring information on its extended brief-term memory neural local area is utilized for the expectation of stock market qualities.

1.6.10 Clinical Prognosis

In clinical science, AI is utilized for disorders analyze. With this, clinical time is growing exceptionally quick and ready to fabricate 3D models that could anticipate the specific capacity of sores inside the cerebrum.

It works with in finding mind growths and other psyche-related sicknesses without issues.

1.6.11 Computerized Language Translation

Nowadays, assuming that we go to a spic-and-span spot and we are not aware of the language, then it is not dependably a difficulty by any means; with respect to this, moreover, contraption acquiring information empowers us via changing the printed content into our respected dialects. Google's GNMT (Google Neural Machine Translation) manages the cost of this element, which is a neural contraption acquiring information that makes an interpretation of the printed content into our familiar language, and it is far known as mechanized interpretation.

The innovation behind the computerized interpretation is a series of acquiring information on calculation that is utilized with photograph acknowledgment and deciphers the text-based substance from one language to some other language [3].

1.6.12 Online Media Features

Online media frameworks use AI calculations and methodologies to make a couple of appealing and extraordinary capacities; for example, facebook notification and data that you exercises; visits, preferences, and criticism; and the time you spend on interesting assortments of posts. Framework acquiring information learns out at your own personal level and makes companions and page hints in your profile.

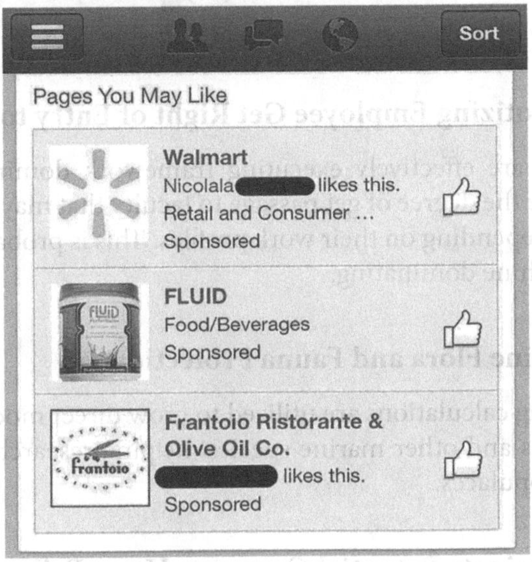

1.6.13 Feeling Evaluation

Feeling appraisal is one of the most outrageous huge tasks of gadget getting to know. Feeling examination is a continuous framework getting to know programming that decides the inclination or assessment of the speaker or the author. As an occasion, assuming that somebody has composed an audit or electronic mail (or any state of a record), a feeling analyzer will promptly find the genuine idea and tone of the text. This feeling assessment utility might be utilized to explore an assessment based absolutely site, choice making bundles, and numerous others.

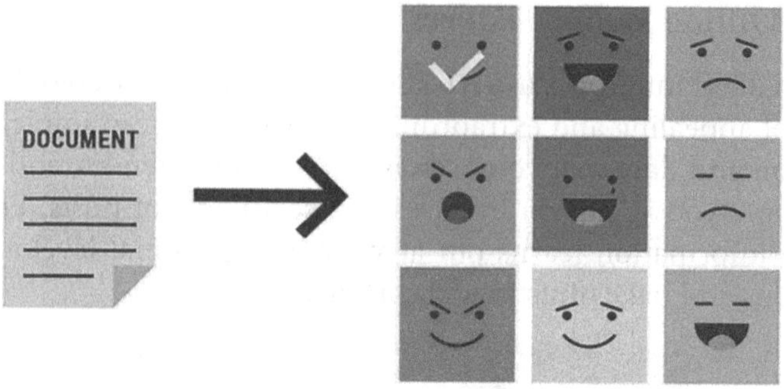

1.6.14 Robotizing Employee Get Right of Entry to Manipulate

Organizations are effectively executing framework dominating calculations to choose the degree of get passage to faculty that may require in various regions, depending on their work profiles. This is probably the coolest bundle of machine dominating.

1.6.15 Marine Flora and Fauna Protection

Gadget learning calculations are utilized to grow direct models for jeopardized cetaceans and other marine species, helping researchers adjust and screen their populaces.

1.6.16 Anticipate Potential Coronary Heart Failure

An arrangement of rules intended to test a wellbeing specialist's freestyle e-notes and see designs in a patient's cardiovascular history is causing ripple effects in cure. Rather than a specialist burrowing through more than one wellbeing data to show up at a genuine analysis, overt repetitiveness is presently diminished with PCs making an assessment in view of to be had information.

1.6.17 Directing Healthcare Efficiency and Scientific Offerings

Colossal medical services areas are effectively looking at the utilization of contraption concentrating on calculations to control better. They anticipate the holding up occasions of patients in the crisis lounge areas all through different divisions of clinics. The models utilize basic components that assist with illustrating the calculation, subtleties of group of laborers at various occasions of day, data of victims, whole logs of branch visits, and the format of trauma centers. Gadget learning calculations, moreover, come to play while recognizing an ailment, treatment arranging, and expectation of the affliction situation. That is one of the greatest imperative gadget acquiring information on applications [1].

1.6.18 Transportation and Commuting (Uber)

At the point when you have utilized an application to book a taxi, you are, as of now, utilizing the device that figures out how much its amount. It gives a redid programming that is specific to you. Mechanically, it identifies your area and offers options to either pass home or working environment or some other continuous locale situated in your records and examples.

It utilizes a gadget dominating arrangement of rules layered on top of verifiable experience data to make an additional a precise ETA forecast. With the execution of gadget considering, they saw a 26% exactness in transportation and Pickup [4].

1.6.19 Dynamic Pricing

Setting the legitimate expense for a decent or supplier is a vintage issue in monetary thought. There are a tremendous amount of evaluating systems that depend upon the objective looked for. Be it a film ticket, an airplane sticker price, or taxi admissions, the entire part is progressively evaluated. In current years, man-made intellectual ability has enabled assessing answers to music buying affinities and conclude extra intense thing costs.

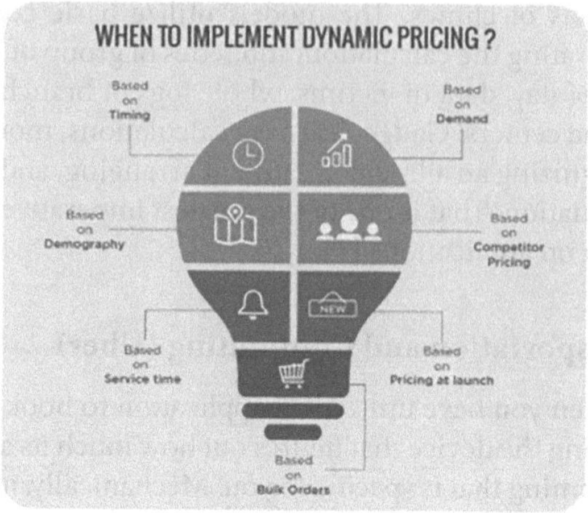

1.6.19.1 How Does Uber Decide the Cost of Your Excursion?

Uber's greatest utilization of framework concentration comes as flood evaluating, a gadget learning model nicknamed "George." In the event that you are getting late for a get-together and you want to digital book a Uber in a packed spot, get prepared to pay twice the ordinary admission. In any event, for flights, on the off chance that you are visiting inside the happy season, the probabilities are that the costs may be twice the interesting charge.

1.6.20 Online Video Streaming (Netflix)

With more than 100,000,000 supporters, doubtlessly that Netflix is the dad of the net streaming worldwide. Netflix's rapid vertical push has all film industrialists extraordinarily astonished, compelling them to inquire, "How in the world would one unmarried web be able to website take on Hollywood?" The arrangement is contraption getting to know.

The Netflix calculation continually assembles immense amounts of data about clients' games like:

- While you stop, rewind, or fast forward
- How day treat watch content material (television proposes on weekdays and movies on weekends)
- The date and time you observe
- While you delay and leave content material (and on the off chance that you at any point return lower)

- The rankings given (around 4 million every day), searches (around 3,000,000 predictable with day)
- Perusing and scrolling conduct

They gather these records for each endorser; they have and utilize their recommender machine and many devices concentrating on programs. That is the reason why they have such a monstrous client maintenance charge [4].

1.7 Challenges in Machine Learning

Machine learning is a powerful and rapidly evolving field but comes with its fair share of challenges. Some of the critical challenges in machine learning include the following.

Data Quality and Quantity: The quality and quantity of data used to train machine learning models greatly influence their performance. Dirty or incomplete data can lead to inaccurate or biased models. Gathering and preprocessing large and diverse datasets can be time-consuming and resource-intensive.
Bias and Fairness: Machine learning models can inadvertently inherit biases in the training data, leading to unfair or discriminatory outcomes. Ensuring fairness and mitigating algorithm bias is a significant challenge, especially in critical applications like hiring, lending, and law enforcement.
Interpretability: Many advanced machine learning models, such as deep neural networks, are often considered "black boxes" because they lack transparency in explaining their decision-making process. This lack of interpretability can hinder trust and regulatory compliance.
Overfitting and Generalization: Overfitting occurs when a model performs well on the training data but poorly on new, unseen data. Striking the right

balance between fitting and generalizing the training data to new data is a continual challenge.

Lack of Data Labeling: Supervised learning relies on labeled data for training, but labeling data can be expensive and time-consuming. Developing methods to label or leverage unlabeled data efficiently is an ongoing challenge.

Scalability and Computational Resources: Training complex machine learning models, intense neural networks, often requires significant computational resources and time. Scaling models to handle larger datasets and deploying them in resource-constrained environments can be challenging.

Hyperparameter Tuning: Selecting appropriate hyperparameters (such as learning rates, batch sizes, and layer sizes) is essential for model performance. However, finding the optimal set of hyperparameters can be a trial-and-error process.

Transfer Learning and Domain Adaptation: Adapting a model trained on one domain to perform well in a different but related domain is challenging. Transfer learning techniques aim to leverage knowledge gained from one task to improve performance on another task.

Privacy and Security: As machine learning is applied to sensitive data, ensuring privacy and security becomes critical. Techniques like differential privacy aim to protect individual privacy while extracting valuable data insights.

Explainability and Accountability: As machine learning is integrated into decision-making processes in various sectors, there is a growing demand for models to explain their predictions. Ensuring accountability and transparency in AI-driven decisions is an ongoing challenge.

Concept Drift and Model Maintenance: Real-world data distributions and relationships can change over time, causing machine learning models to become less effective. Continuous monitoring and retraining of models to adapt to evolving data are essential.

Human–Machine Collaboration: Determining the appropriate roles for humans and machines in complex tasks and designing effective interfaces for collaboration poses challenges.

Addressing these challenges requires collaboration among researchers, practitioners, policymakers, and ethicists to ensure that machine learning technologies are developed, deployed, and regulated responsibly and ethically.

1.8 Limitations of Machine Learning

While a powerful and versatile technology, machine learning also has certain limitations that can impact its performance and application. Some of the critical limitations of machine learning include the following.

Data Dependence: Machine learning models rely heavily on the training data quality, quantity, and relevance. The model's performance can be compromised if the data are biased, incomplete, or not representative of the real-world scenario.

Limited Generalization: While machine learning models can generalize patterns from training data to make predictions on new data, they might struggle when faced with data that differs significantly from what they were trained on. Generalization performance can suffer when new data distribution deviates from the training data.

Interpretable Outputs: Many complex machine learning models, and intense neural networks, produce difficult results to interpret or explain. This lack of transparency can be problematic in critical applications where understanding the reasoning behind a decision is essential.

Need for Large Amounts of Data: Deep learning models, in particular, often require massive amounts of data to achieve high performance. Gathering and labeling such datasets can be time-consuming, expensive, or even infeasible in some cases.

High Computational Demands: Training sophisticated machine learning models, intense neural networks, demands substantial computational resources, including powerful GPUs and significant memory. This can limit the accessibility of these methods to those with sufficient computational infrastructure.

Curse of Dimensionality: As the number of features (dimensions) in the data increases, the amount of data required to train a model effectively grows exponentially. This "curse of dimensionality" can challenge training accurate models in high-dimensional spaces.

Overfitting and Underfitting: Balancing the complexity of a model is essential to prevent overfitting (fitting noise in the data) or underfitting (insufficiently capturing patterns). Achieving this balance often requires careful tuning of model hyperparameters.

Limited Transfer Learning: While transfer learning allows models to leverage knowledge from one domain to improve performance in another, it

may not always work seamlessly. Domain differences or shifts can hinder the effectiveness of transfer learning.
Lack of Common Sense and Context: Machine learning models lack the common sense reasoning and contextual understanding that humans possess. They may struggle with tasks requiring contextual understanding, reasoning, or ambiguity.
Ethical and Bias Concerns: Machine learning models can inherit biases in the training data, leading to unfair or discriminatory outcomes. Ensuring fairness and mitigating bias is challenging, particularly in sensitive applications like healthcare and criminal justice.
Unpredictable Errors: Machine learning models can make unexpected errors that are difficult to anticipate or diagnose. These errors may occur in novel or adversarial scenarios on which the model was not explicitly trained.
Continuous Learning and Adaptation: While machine learning models can be trained initially, adapting them to changing environments or evolving data distributions can be challenging. Maintaining model accuracy over time requires ongoing monitoring and updates.
Human–Machine Collaboration: Fully automating complex tasks may not always be feasible or desirable. Determining the optimal balance between human expertise and machine assistance can be challenging.
Security and Privacy Concerns: Adversarial attacks can exploit vulnerabilities in machine learning models, potentially causing them to produce incorrect outputs. Privacy concerns arise when models handle sensitive or personal data.
Energy Consumption: Training and running complex machine learning models can consume significant energy, contributing to environmental concerns.

While these limitations highlight the current challenges in machine learning, ongoing research and innovation aim to address these issues and further improve the capabilities of machine learning technologies.

1.9 Projects in Machine Learning

There have been numerous important and influential projects in the field of machine learning that have advanced our understanding and capabilities in various domains. Here are some notable projects that have had a significant impact.

ImageNet and AlexNet: The ImageNet Large-Scale Visual Recognition Challenge spurred the development of deep convolutional neural networks (CNNs) for image classification. AlexNet, introduced in 2012, was a pioneering deep CNN architecture that achieved a breakthrough in image classification accuracy [5].
AlphaGo: Developed by DeepMind, AlphaGo demonstrated remarkable proficiency in playing the ancient board game Go, defeating world champion Lee Sedol in 2016. AlphaGo's success showcased the power of deep reinforcement learning and led to advancements in AI research.
BERT (Bidirectional Encoder Representations from Transformers): BERT, introduced by Google AI in 2018, revolutionized natural language processing by training transformers to understand the context from both directions in a sentence. It significantly improved performance on various NLP tasks [6].
GPT (Generative Pre-trained Transformer) Series: The GPT series, including GPT-2 and GPT-3, developed by OpenAI, pushed the boundaries of language generation and understanding. GPT-3, in particular, demonstrated the potential of large-scale language models for diverse applications.
YOLO (You Only Look Once): YOLO is an object detection algorithm that can detect and classify objects in images and video frames in real-time. Its speed and accuracy have made it a popular choice for computer vision applications.
ResNet (Residual Networks): ResNet introduced the concept of residual learning, enabling the training of intense neural networks. This architecture has been crucial for advancing the performance of deep networks in image recognition tasks.
StyleGAN: StyleGAN, developed by NVIDIA, allows for generating highly realistic and customizable images using generative adversarial networks (GANs). It has been used for creating art, animation, and even generating human-like faces.
Transformers: The transformer architecture, introduced in the "Attention Is All You Need" paper, has become a cornerstone of various natural language processing tasks. It has led to advancements in machine translation, text generation, and more.

DeepMind's Atari and OpenAI's Gym: These projects introduced benchmark environments for testing and developing reinforcement learning algorithms. They contributed to advancements in deep reinforcement learning and agent performance on complex tasks.

Deep Dream: Developed by Google, Deep Dream is a visualization technique that produces artistic images by enhancing and manipulating neural network activations. It showcases the creative potential of neural networks beyond their original purposes.
COVID-19 Research: Machine learning played a crucial role in analyzing and predicting the spread of COVID-19, identifying potential drug candidates, and understanding the impact of the pandemic. Various research projects utilized machine learning to contribute to pandemic-related insights.
Waymo (formerly Google Self-Driving Car Project): Waymo's self-driving car project, now an independent company under Alphabet, has leveraged machine learning to develop autonomous vehicles capable of navigating complex real-world environments.
These projects represent just a subset of the influential work in machine learning. The field is dynamic, with ongoing research and projects constantly pushing the boundaries of what is possible and driving innovation across various domains.

References

1. https://www.simplilearn.com/tutorials/machine-learning-tutorial/machine-learning-applications
2. https://www.upgrad.com/blog/machine-learning-and-its-breakthrough-applications/
3. https://www.javatpoint.com/applications-of-machine-learning\
4. https://www.edureka.co/blog/machine-learning-applications/
5. https://data-flair.training/blogs/stock-price-prediction-machine-learning-project-in-python/
6. https://data-flair.training/blogs/r-data-science-project-uber-data-analysis/

2

Machine Learning Building Blocks

2.1 Data Collection

The first stage of the machine learning (ML) life cycle is data gathering. This step's objective is to locate and collect all data-related issues.

As data can be gathered from a variety of sources, including files, databases, the internet, and mobile devices, we must first identify the various data sources in this stage. One of the most crucial phases of the life cycle is data gathering. The effectiveness of the output will depend on the quantity and calibre of the data gathered. The prediction will be more accurate the more data there is.

2.1.1 Importing the Data from CSV Files

The pandas package contains a function that is frequently used for importing datasets. You can import data in a variety of formats, including csv files and xlsx, among others. We will employ the exact same function. To import the dataset and print the first 10 rows of the data, use the code below. Before reading the data, we will import the pandas package.

2.2 Data Preparation

We must prepare the data for future use after gathering it. Data preparation is the process of organizing and preparing our data for use in ML training.

In this stage, we initially group all the data together before randomly arranging them.

This method may be separated into two different steps.

Kolla Bhanu Prakash. Machine Learning for Industrial Applications, (21–50) © 2024 Scrivener Publishing LLC

2.2.1 Data Exploration

It helps us comprehend the type of data that we must work with. We must comprehend the qualities, formats, and properties of the data.

A more accurate grasp of the data results in successful results. We discover correlations, broad patterns, and outliers in this.

In the process of exploratory data analysis (EDA), we look into the dataset to become comfortable with it. We learn about the data structure, missing values, type, etc. Before creating an ML model, all operations that are performed on data fall under the category of EDA.

2.2.2 Data Pre-Processing

Data pre-processing is the first and crucial step while creating an ML model. Data pre-processing tasks are the required tasks for cleaning the data and making it suitable for an ML model. It also increases the accuracy and efficiency of the ML models that are created.

Some specified ML model needs information in a specified format, for example, random forest algorithm does not support null values; therefore, to execute random forest algorithm, null values have to be managed from original raw dataset. Dataset should be formatted in such a way that more than one machine learning and deep learning algorithm are executed in one dataset, and the best out of them is chosen.

2.3 Data Wrangling

It is becoming more and more crucial to organize the correct data for analysis as the world of data is expanding so quickly. Practically, every company decision is based on data and information used by business users. Making raw data accessible for analytics is crucial. Data transformation and mapping are steps in the process of preparing raw data for analysis.

Cleaning, organizing, and manipulating raw data is referred to as data wrangling. It allows companies to work with more complex data more quickly. It is also known as data cleaning or data munging. Depending on your data and the objective you are trying to attain, different projects will use different precise methodologies.

An essential component of data processing is data wrangling software. With only 20% of their time left over for exploration and modeling, data professionals spend roughly 80% of their time manipulating the data. Data mining entails obtaining data and making sense of it; data wrangling serves as a first step in this process.

2.4 Data Analysis

Today's businesses have less margins for mistake. They must gather as much pertinent, relevant data as they can and utilize it to guide their judgments.

Data analysis is the process of looking through raw data to find relevant, useful information that can be used to inform business decisions. The process reduces decision-making risks by offering insightful information and statistics.

Here is a list of the key factors that make data analysis so important in today's corporate world.

1. Data analysis enables you to determine where to concentrate your advertising efforts.
2. Your company can learn more about the purchasing patterns, disposable income, and most likely interests of your target market through data research.
3. Data offers information to businesses.
4. "You Get More Reliable Information Data" is necessary if you want to make educated decisions, but there are other factors as well.

2.5 Model Selection

In the context of ML, the term "model selection" can be interpreted in a variety of ways that reflect various degrees of abstraction. One reason is that we might want to choose the optimum hyperparameters for a certain ML approach. The parameters of the learning process itself that must be specified a priori, or before model fitting, are known as hyperparameters. On the other hand, model parameters are parameters that derive from the fit.

For instance, in a logistic regression model, the regularization type and strength (if any) are hyperparameters that must be specified before fitting, whereas the fitted model's coefficients are model parameters. For a model to perform well on a set of data, choosing the appropriate hyperparameters can be essential.

Another reason is that we might want to choose the best learning method (and its corresponding "optimal" hyperparameters) from a list of available ML techniques. Hereafter, we will refer to this as algorithm

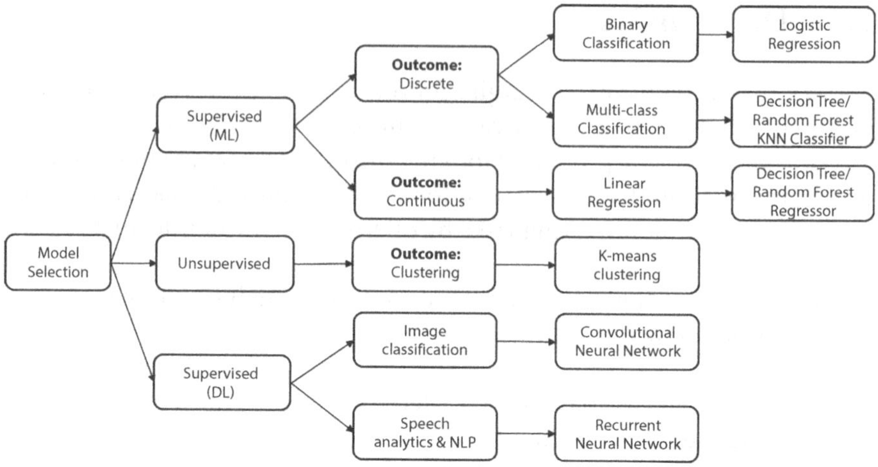

Figure 2.1 Model selection.

selection. We might consider the classification problem at hand and wonder, for example, whether a logistic regression model or a random forest classifier produces the best classification performance on the assigned task.

2.6 Model Building

Building an ML model with the intention of applying a variety of analytical methods to examine the data and evaluate the results is the goal of this phase. The process begins with identifying the problem type, after which we choose from a variety of ML approaches, including classification, regression, cluster analysis, association, and others. We then use the data that has been prepared to develop the model, which is then evaluated.

Data must be divided into two sets, namely, a "training set" and a "testing set," at a ratio of 80:20 or 70:30, in order to build an ML model. Depending on the type of input data and the desired business outcome to be predicted, a variety of supervised (for labeled data) and unsupervised (for unlabeled data) methods are available. If the outcome to predict is binary (0/1) in nature, then it is advised to use the logistic regression algorithm. If the outcome to predict is multi-class (1/2/3/...) in nature, then it is advised to use the decision tree classifier, random forest classifier, or k-nearest neighbor (KNN) algorithm. If the outcome to predict is continuous in nature, then it is advised to use the linear regression algorithm (e.g., decision tree regressor and random forest regressor). When analyzing unlabeled or unstructured (text) data, the clustering (unsupervised) algorithm is preferred. Other unstructured (image/voice) data types are suggested for analysis using artificial neural network algorithms, such as convolutional neural networks (CNN) for image recognition and recurrent neural networks (RNN) for voice recognition and natural language processing. The training dataset is used to create the model, and the test dataset is used to make predictions. For better performance, deep learning (neural networks) models are preferred over regression models (ML models), as these models introduce an additional layer of non-linearity with the introduction of activation function.

2.7 Model Evaluation

The most crucial thing you can do to properly evaluate your model is to avoid training it on the whole dataset. Typically, 70% of the data would be used for training and 30% for testing.

To avoid the possibility of overfitting to the training set, it is crucial to use fresh data while assessing our model. However, there are times when it is beneficial to evaluate our model as we build it to identify the best model parameters. However, we cannot use the test set for this evaluation because otherwise, we risk choosing the parameters that perform best on the test data rather than the parameters that generalize the best. We produce a third subset of the data, called the validation set, to test the model while it is still being built and tuned. Sixty percent of the data would be used for training, 20% for validation, and 20% for testing, according to a usual train, test, and validation split.

2.7.1 Classification Metrics

There are four possible outcomes when making classification predictions.

1. When you correctly anticipate that an observation belongs to a class and it actually does, you have a **true positive**.
2. When you forecast that an observation does not belong to a class and it truly does not belong to that class, you have a **true negative.**
3. When you forecast that an observation belongs to a class when it does not, you get **false positives.**
4. When you forecast that an observation does not belong to a class when it actually does, you get **false negatives.**

Accuracy, precision, and recall are the three primary measures used to gauge the effectiveness of a classification algorithm.

2.7.1.1 *Accuracy*

The percentage of accurate predictions for the test data is what is meant by accuracy. By dividing the number of accurate predictions by the total number of predictions, it may be simply determined.

$$\text{accuracy} = \frac{\text{correct predictions}}{\text{all predictions}}$$

2.7.1.2 *Precision*

Precision is defined as the percentage of instances that are truly relevant (also known as true positives) out of all the examples that were predicted to fall into a particular class.

$$\text{precision} = \frac{\text{true positives}}{\text{true positives} + \text{false positives}}$$

2.7.1.3 *Recall*

The percentage of examples that were correctly identified as belonging to a class in relation to all of the examples in the class is known as recall.

$$\text{recall} = \frac{\text{true positives}}{\text{true positives} + \text{false negatives}}$$

2.7.2 Regression Metrics

The following three error metrics are frequently used for assessing and reporting the effectiveness of a regression model.

2.7.2.1 *Mean Squared Error*

The average of the squared differences between the projected output and the actual output is how mean squared error is commonly referred to. It is commonly used because it does not differentiate between whether the prediction was too high or too low, and squared error simply indicates that the prediction was inaccurate.

$$MSE(y_{true}, y_{pred}) = \frac{1}{n_{samples}} \sum (y_{true} - y_{pred})^2$$

2.7.2.2 *Root Mean Squared Error*

One of the methods most frequently used to assess the accuracy of forecasts is root mean square error, also known as root mean square deviation. It illustrates the Euclidean distance between measured true values and forecasts.

Calculate the residual (difference between prediction and truth) for each data point, along with its norm, mean, and square root in order to determine the root mean square error (RMSE). Due to the fact that it requires and uses real measurements at each projected data point, RMSE is frequently utilised in supervised learning applications.

$$RMSE = \sqrt{\frac{\sum_{i=1}^{N} ||y(i) - \hat{y}(i)||^2}{N}},$$

2.7.2.3 *Mean Absolute Error*

Absolute error in the context of ML refers to the size of the discrepancy between the forecast of an observation and its actual value. The size of errors for the entire group is determined by MAE by averaging the absolute errors for a set of forecasts and observations. MAE is another name for L1 loss function.

$$\text{MAE} = \frac{|(y_i - y_p)|}{n}$$

y_i = actual value
y_p = predicted value
n = number of observations/rows

2.8 Deployment

There are various methods for deploying ML models into production, storing them, and managing predictive models after deployment. Selecting the best strategy for a use case can be difficult, but, with the team's technical and analytical maturity, the overall organization structure and its interactions can assist in making this decision.

The process of integrating an ML model into an already existing production environment is known as deployment, and it allows you to use data to make useful business decisions. It can be one of the most challenging stages of the ML life cycle and is one of the final ones. Frequently, traditional model-building languages are incompatible with an organization's IT systems, requiring data scientists and programmers to spend valuable time and brainpower rewriting them.

We can deploy the machine learning model in three ways:

1. Deploying ML models as web services.
2. Deploying ML models for batch prediction.
3. Deploying ML models on edge devices as embedded models.

2.8.1 Machine Learning Projects

In ML, as a beginner, we start learning with the well-known project like detecting spam/not spam.

2.8.2 Spam Detection Using Machine Learning

Before starting the project, it is important to understand what spam actually is. Spam is any unwanted, uninvited digital communication, frequently an email that is distributed in large quantities. Spam is a major time and money waster. Spam, which includes credit card offers, discount offers, traffic plans, and promotions, can be found in emails, YouTube comments, and SMS messages.

Mostly, we can see in our Gmail that new emails are automatically classified as essential, common, and spam. Spam emails land in our spam box, and important emails with the crucial symbol land in our inbox.

Multi-layer perceptron, decision tree, and naïve Bayes classifier are three of the most widely used ML algorithms for spam filtering, along with the use of natural language processing.

2.8.3 Spam Detection for YouTube Comments Using Naïve Bayes Classifier

Dataset: We have used the dataset from UCI Machine Learning Repository that provides the YouTube Spam Collection Dataset and contains five attributes and number of instances that are 1956.

Dataset --- YouTube ID -- # Spam - # Ham - Total
Psy ------- 9bZkp7q19f0 --- 175 --- 175 --- 350
KatyPerry - CevxZvSJLk8 --- 175 --- 175 --- 350
LMFAO ----- KQ6zr6kCPj8 --- 236 --- 202 --- 438
Eminem ---- uelHwf8o7_U --- 245 --- 203 --- 448
Shakira --- pRpeEdMmmQ0 --- 174 --- 196 --- 370

Code:

```
import pandas as pd

import numpy as np
from sklearn.metrics import confusion_matrix, classification_report
Psy = pd.read_csv("Youtube01-Psy.csv")
Katy = pd.read_csv("Youtube02-KatyPerry.csv")
LMFAO = pd.read_csv("Youtube03-LMFAO.csv")
Eminem = pd.read_csv("Youtube04-Eminem.csv")
Shakira = pd.read_csv("Youtube05-Shakira.csv")
Psy.head()
```

	COMMENT_ID	AUTHOR	DATE	CONTENT	CLASS
0	LZQPQhLyRh80UYxNuaDWhIGQYNQ96IuCg-AYWqNPjpU	Julius NM	2013-11-07T06:20:48	Huh, anyway check out this you[tube] channel: ...	1
1	LZQPQhLyRh_C2cTtd9MvFRJedxydaVW-2sNg5Diuo4A	adam riyati	2013-11-07T12:37:15	Hey guys check out my new channel and our firs...	1
2	LZQPQhLyRh9MSZYnf8djyk0gEF9BHDPYrrK-qCczIY8	Evgeny Murashkin	2013-11-08T17:34:21	just for test I have to say murdev.com	1
3	z13jhp0bxqncu512g22wvzkasxmvvzjaz04	ElNino Melendez	2013-11-09T08:28:43	me shaking my sexy ass on my channel enjoy ^_^	1
4	z13fwbwp1oujthgqj04chlngpvzmtt3r3dw	GsMega	2013-11-10T16:05:38	watch?v=vtaRGgvGtWQ Check this out .	1

Figure 2.2 Data sample of YouTube comments of Psy.

Katy.head()

	COMMENT_ID	AUTHOR	DATE	CONTENT	CLASS
0	z12pgdhovmrktzm3i23es5d5junftft3f	lekanaVEVO1	2014-07-22T15:27:50	i love this so much. AND also I Generate Free ...	1
1	z13yx345uxepetggz04ci5rjcxeohzlrtf4	Pyunghee	2014-07-27T01:57:16	http://www.billboard.com/articles/columns/pop-...	1
2	z12lsjvi3wa5x1vwh04cibeaqnzrevxajw00k	Erica Ross	2014-07-27T02:51:43	Hey guys! Please join me in my fight to help a...	1
3	z13jcjuovxbwfr0ge04cev2ipsjdfdurwck	Aviel Haimov	2014-08-01T12:27:48	http://psnboss.com/?ref=2tGgp3pV6L this is the...	1
4	z13qybua2yfydzxzj04cgfpqdt2syfx53ms0k	John Bello	2014-08-01T21:04:03	Hey everyone. Watch this trailer!!!!!!!! http...	1

Figure 2.3 Data sample of YouTube comments of Katy.

LMFAO.head()

	COMMENT_ID	AUTHOR	DATE	CONTENT	CLASS
0	z13uwn2heqndtr5g304ccv5j5kqqzxjadmc0k	Corey Wilson	2015-05-28T21:39:52.376000	<a href="http://www.youtube.com/watch?v=KQ6zr6...	0
1	z124jvczaz3dxhnbc04cffk43oiugj25yzo0k	Epic Gaming	2015-05-28T20:07:20.610000	wierd but funny	0
2	z13tczjy5xj0vjmu5231unho1ofey5zdk	LaS Music	2015-05-28T19:23:35.355000	Hey guys, I'm a human. Bu...	1
3	z13tzr0hdpnayhqqc04cd3zqqqjkf3ngckk0k	Cheryl Fox	2015-05-28T17:49:35.294000	Party Rock....lol...who wants to shuffle!!!	0
4	z12pcvix4zedcjvyb04ccr1r0mr2g5xwyng0k	PATRICK_TW	2015-05-28T16:28:26.818000	Party rock	0

Figure 2.4 Data sample of YouTube comments of LMFAO.

Eminem.head()

	COMMENT_ID	AUTHOR	DATE	CONTENT	CLASS
0	z12rwfnyyrbsefonb232i5ehdxzkjzjs2	Lisa Wellas	NaN	+447935454150 lovely girl talk to me xxx	1
1	z130wpnwwnyuetxcn23xf5k5ynmkdpjrj04	jason graham	2015-05-29T02:26:10.652000	I always end up coming back to this song 	0
2	z13vsfqirtavjvu0t22ezrgzyorwxhpf3	Ajkal Khan	NaN	my sister just received over 6,500 new <a rel=...	1
3	z12wjzc4eprnvja4304cgbbizuved35wxcs	Dakota Taylor	2015-05-29T02:13:07.810000	Cool	0
4	z13xjfr42z3uxdz2223gx5rrzs3dt5hna	Jihad Naser	NaN	Hello I'am from Palastine	1

Figure 2.5 Data sample of YouTube cmments of Eminem.

Shakira.head()

	COMMENT_ID	AUTHOR	DATE	CONTENT	CLASS
0	z13lgffb5w3ddx1ul22qy1wxspy5cpkz504	dharma pal	2015-05-29T02:30:18.971000	Nice song	0
1	z123dbgb0mqjfxbtz22ucjc5jvzcv3ykj	Tiza Arellano	2015-05-29T00:14:48.748000	I love song	0
2	z12quxxp2vutflkxv04cihggzt2azl34pms0k	Priñçeśś Åliś Łøvê Dømîñø Mâđiś™	2015-05-28T21:00:08.607000	I love song	0
3	z12icv3ysqvlwth2c23eddlykyqut5z1h	Eric Gonzalez	2015-05-28T20:47:12.193000	860,000,000 lets make it first female to reach...	0
4	z133stly3kete3tly22petvwdpmghrlli	Analena López	2015-05-28T17:08:29.827000	shakira is best for worldcup	0

Figure 2.6 Data sample of YouTube comments of Shakira.

```
# Using pandas concat function to combine all the 5 csv files as one dataset.
data = pd.concat([Psy, Katy, LMFAO, Eminem, Shakira])
data.head()
```

	COMMENT_ID	AUTHOR	DATE	CONTENT	CLASS
0	LZQPQhLyRh80UYxNuaDWhIGQYNQ96IuCg-AYWqNPjpU	Julius NM	2013-11-07T06:20:48	Huh, anyway check out this you[tube] channel: ...	1
1	LZQPQhLyRh_C2cTtd9MvFRJedxydaVW-2sNg5Diuo4A	adam riyati	2013-11-07T12:37:15	Hey guys check out my new channel and our firs...	1
2	LZQPQhLyRh9MSZYnf8djyk0gEF9BHDPYrrK-qCczIY8	Evgeny Murashkin	2013-11-08T17:34:21	just for test I have to say murdev.com	1
3	z13jhp0bxqncu512g22wvzkasxmvvzjaz04	ElNino Melendez	2013-11-09T08:28:43	me shaking my sexy ass on my channel enjoy ^_^	1
4	z13fwbwp1oujthgqj04chlngpvzmtt3r3dw	GsMega	2013-11-10T16:05:38	watch?v=vtaRGgvGtWQ Check this out .	1

Figure 2.7 Data sample of concatenated YouTube comments.

```
data.shape
```

```
(1956, 5)
```

```
# here we are droping the unnecessary columns
data.drop(["COMMENT_ID","DATE","AUTHOR"],axis=1,inplace=True)
data.info()
```

```
<class 'pandas.core.frame.DataFrame'>
Int64Index: 1956 entries, 0 to 369
Data columns (total 2 columns):
 #   Column   Non-Null Count  Dtype
---  ------   --------------  -----
 0   CONTENT  1956 non-null   object
 1   CLASS    1956 non-null   int64
dtypes: int64(1), object(1)
memory usage: 45.8+ KB
```

```
# Here we are splitting our dataset for training and testing
from sklearn.model_selection import train_test_split
X_train, X_test, y_train, y_test = train_test_split(data["CONTENT"],
data["CLASS"])
# Tokenizing comments using Text Vectorization
from sklearn.feature_extraction.text import TfidfVectorizer
tfidf_vector = TfidfVectorizer(use_idf=True, lowercase=True)
X_train_final = tfidf_vector.fit_transform(X_train)
X_train_final.shape
```

```
(1467, 3616)
```

```
from sklearn.naive_bayes import MultinomialNB
model = MultinomialNB()
model.fit(X_train_final, y_train)
```

```
MultinomialNB()
```

```
X_test_final = tfidf_vector.transform(X_test)
predictions = model.predict(X_test_final)
confusion_matrix(y_test, predictions)
```

```
array([[213,  31],
       [ 11, 234]])
```

```
print(classification_report(y_test, predictions))
```

```
              precision    recall  f1-score   support

           0       0.95      0.87      0.91       244
           1       0.88      0.96      0.92       245

    accuracy                           0.91       489
   macro avg       0.92      0.91      0.91       489
weighted avg       0.92      0.91      0.91       489
```

```
model.score(X_test_final, y_test)
```

```
0.9141104294478528
```

2.8.4 Fake News Detection

Fake news, a form of yellow journalism, refers to news items that may be hoaxes and is typically disseminated through social media and other online media. This is frequently accomplished with political objectives in order to advance or impose particular beliefs. Such news stories may make misleading or overstated claims, become viralized by algorithms, and trap users in a filter bubble.

This Python project's dataset, news.csv, will be used for it. 77964 is the shape of this dataset. The title and body of the news are listed in the second and third columns, respectively, and the news' authenticity is indicated in the fourth column by labels in the first three columns.

```
import numpy as np
import pandas as pd
import itertools
from sklearn.model_selection import train_test_split
from sklearn.feature_extraction.text import TfidfVectorizer
from sklearn.linear_model import PassiveAggressiveClassifier
from sklearn.metrics import accuracy_score, confusion_matrix
df=pd.read_csv('news.csv')
df.head()
```

	Unnamed: 0	title	text	label
0	8476	You Can Smell Hillary's Fear	Daniel Greenfield, a Shillman Journalism Fello...	FAKE
1	10294	Watch The Exact Moment Paul Ryan Committed Pol...	Google Pinterest Digg Linkedin Reddit Stumbleu...	FAKE
2	3608	Kerry to go to Paris in gesture of sympathy	U.S. Secretary of State John F. Kerry said Mon...	REAL
3	10142	Bernie supporters on Twitter erupt in anger ag...	— Kaydee King (@KaydeeKing) November 9, 2016 T...	FAKE
4	875	The Battle of New York: Why This Primary Matters	It's primary day in New York and front-runners...	REAL

Figure 2.8 Data sample.

```
df.shape
```

```
(6335, 4)
```

```
labels=df.label
labels.head()
```

```
0    FAKE
1    FAKE
2    REAL
3    FAKE
4    REAL
Name: label, dtype: object
```

```
#Split the dataset
x_train,x_test,y_train,y_test=train_test_split(df['text'], labels, test_size=0.2, random_state=7)
#Initialize a TfidfVectorizer
tfidf_vectorizer=TfidfVectorizer(stop_words='english', max_df=0.7)
#Fit and transform train set, transform test set
tfidf_train=tfidf_vectorizer.fit_transform(x_train)
tfidf_test=tfidf_vectorizer.transform(x_test)
#Initialize a PassiveAggressiveClassifier
pac=PassiveAggressiveClassifier(max_iter=50)
pac.fit(tfidf_train,y_train)
#Predict on the test set and calculate accuracy
y_pred=pac.predict(tfidf_test)
score=accuracy_score(y_test,y_pred)
print(f'Accuracy: {round(score*100,2)}%')
```

```
Accuracy: 92.82%
```

```
#Build confusion matrix
confusion_matrix(y_test,y_pred, labels=['FAKE','REAL'])
```

```
array([[589,  49],
       [ 42, 587]])
```

2.8.5 House Price Prediction

Given a collection of details about a Boston residence, our ML model must be able to forecast the property price. This is the issue that we will try to resolve. Our ML model will be trained on housing data from Boston using the scikit-learn Boston dataset.

The UCI Machine Learning Repository provided the dataset for this project. Each of the 506 entries in this dataset, which was gathered in 1978, includes aggregate data on 14 characteristics of homes in various Boston suburbs.

```
from sklearn.linear_model import LinearRegression
from sklearn.model_selection import train_test_split
from sklearn import metrics
import pandas as pd
import matplotlib.pyplot as plt
%matplotlib inline
import seaborn as sns
df = pd.read_csv("BostonHousing.csv")
df.head()
```

	crim	zn	indus	chas	nox	rm	age	dis	rad	tax	ptratio	b	lstat	medv
0	0.00632	18.0	2.31	0	0.538	6.575	65.2	4.0900	1	296	15.3	396.90	4.98	24.0
1	0.02731	0.0	7.07	0	0.469	6.421	78.9	4.9671	2	242	17.8	396.90	9.14	21.6
2	0.02729	0.0	7.07	0	0.469	7.185	61.1	4.9671	2	242	17.8	392.83	4.03	34.7
3	0.03237	0.0	2.18	0	0.458	6.998	45.8	6.0622	3	222	18.7	394.63	2.94	33.4
4	0.06905	0.0	2.18	0	0.458	7.147	54.2	6.0622	3	222	18.7	396.90	5.33	36.2

```
df.shape
(506, 14)
```

```
df.isnull().sum()
crim       0
zn         0
indus      0
chas       0
nox        0
rm         0
age        0
dis        0
rad        0
tax        0
ptratio    0
b          0
lstat      0
medv       0
dtype: int64
```

Plotting graph between CRIM and MEDV for visualizing the effect of crim rate on house price

```
for i in df.columns[:-1]:
    plt.figure(figsize=(6,4))
    plt.title("{} vs .\nMEDV".format(i),fontsize=16)
    sns.regplot(x=df[i],y=df['medv'])
    plt.grid(True)
    plt.xlabel(i,fontsize=14)
    plt.ylabel('MEDV',fontsize=14)
    plt.show()
```

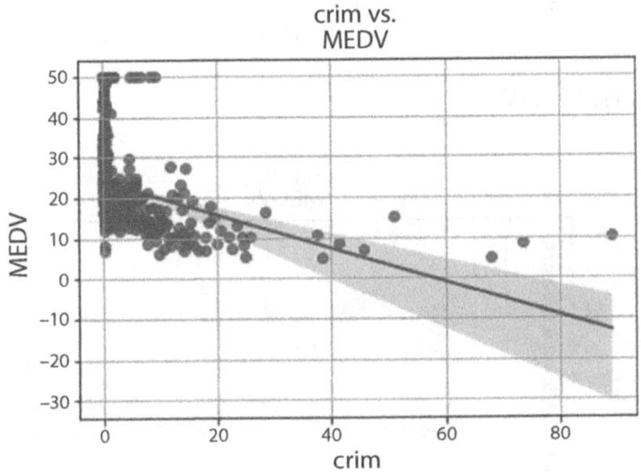

Figure 2.9 Crim vs. MEDV.

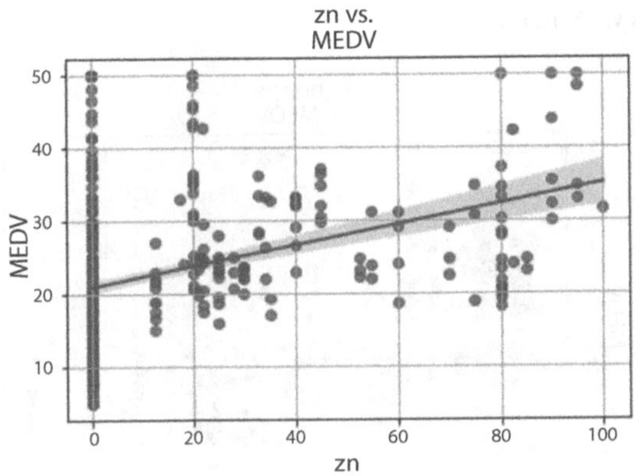

Figure 2.10 Zn vs. MEDV.

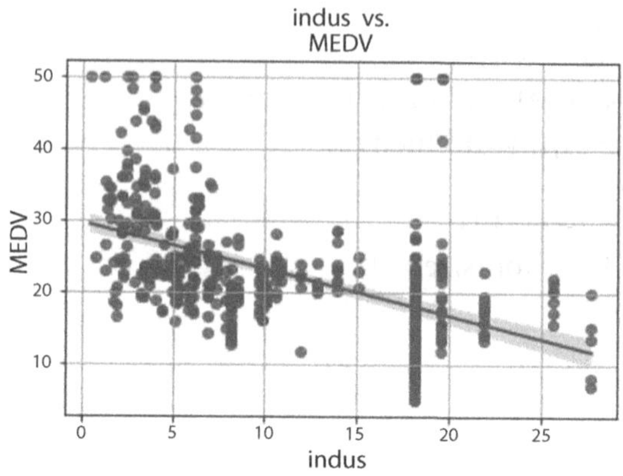

Figure 2.11 Indus vs. MEDV.

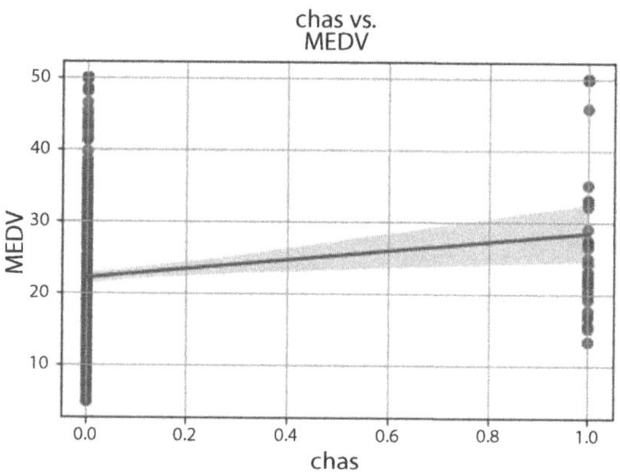

Figure 2.12 Chas vs. MEDV.

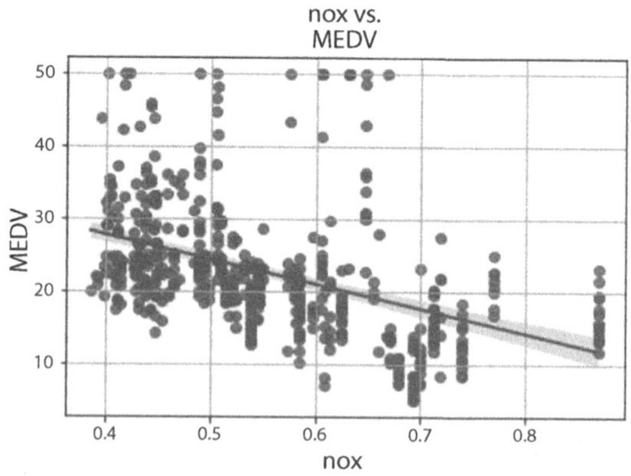

Figure 2.13 Nox vs. MEDV.

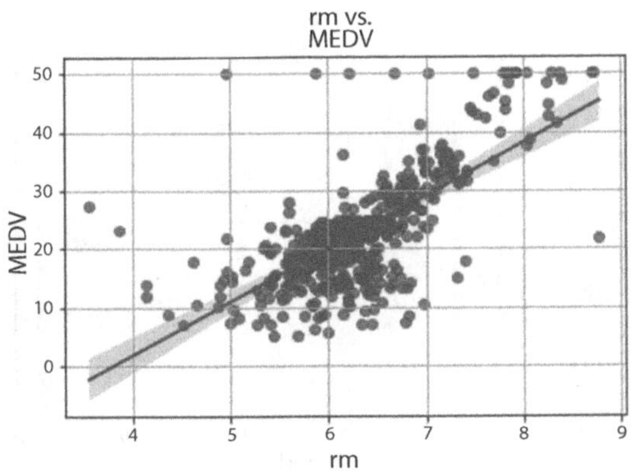

Figure 2.14 Rm vs. MEDV.

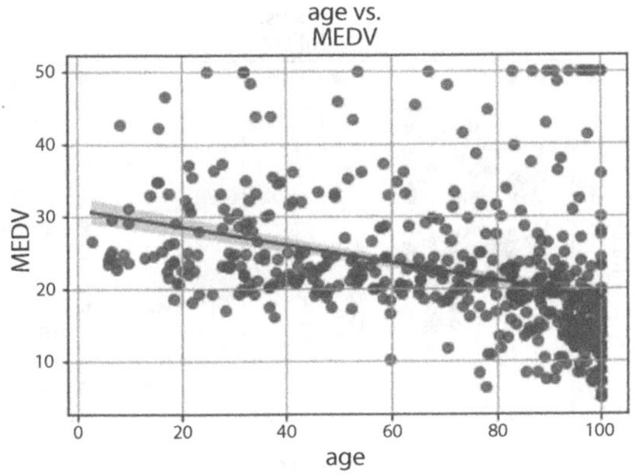

Figure 2.15 Age vs. MEDV.

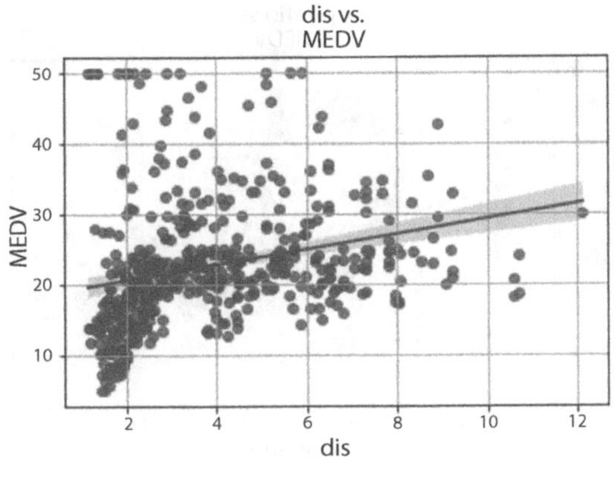

Figure 2.16 Dis vs. MEDV.

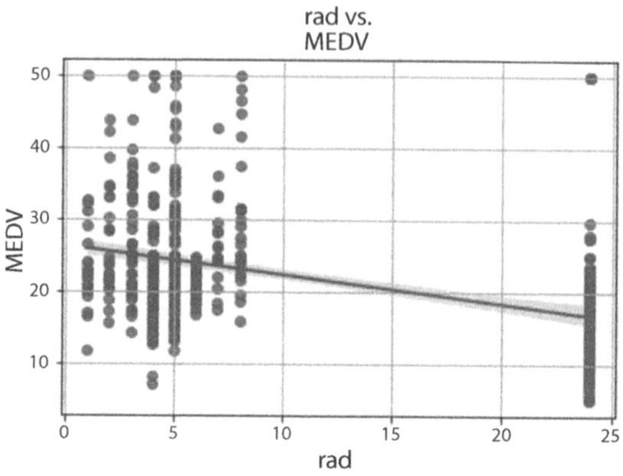

Figure 2.17 Rad vs. MEDV.

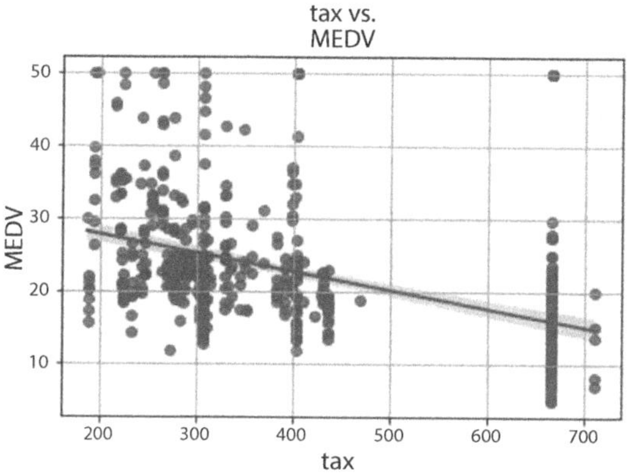

Figure 2.18 Tax vs. MEDV.

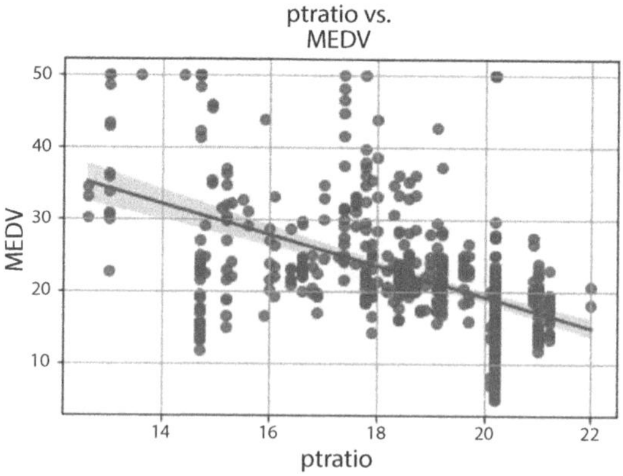

Figure 2.19 Ptratio vs. MEDV.

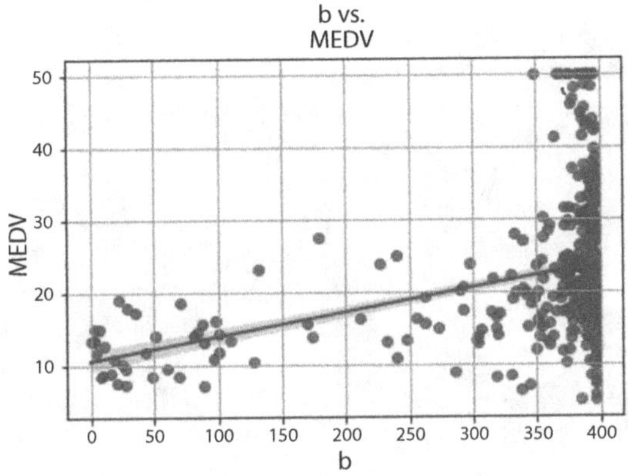

Figure 2.20 B vs. MEDV.

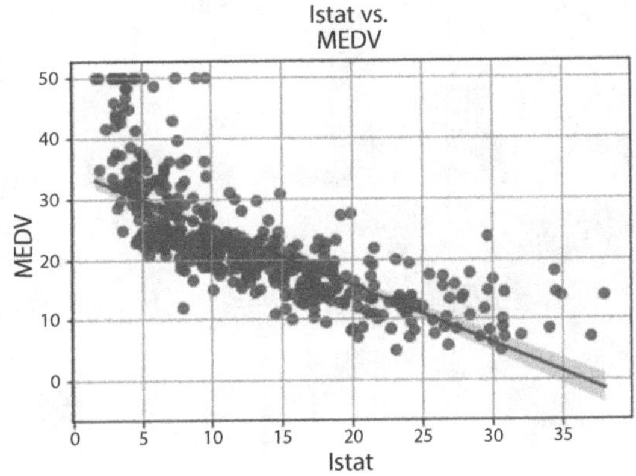

Figure 2.21 lstat vs. MEDV.

```
# Input or independent value
X = boston_data.iloc[:, :13]
# Output or dependent value
Y = boston_data["medv"]
# Find correlation between all available columns
corr = df.corr()
fig, ax = plt.subplots(figsize=(10,10))
# Find correlated data columns using heatmaps
sns.heatmap(corr, annot=True, fmt=".2f")
plt.xticks(range(len(corr.columns)), corr.columns);
plt.yticks(range(len(corr.columns)), corr.columns)
plt.show()
```

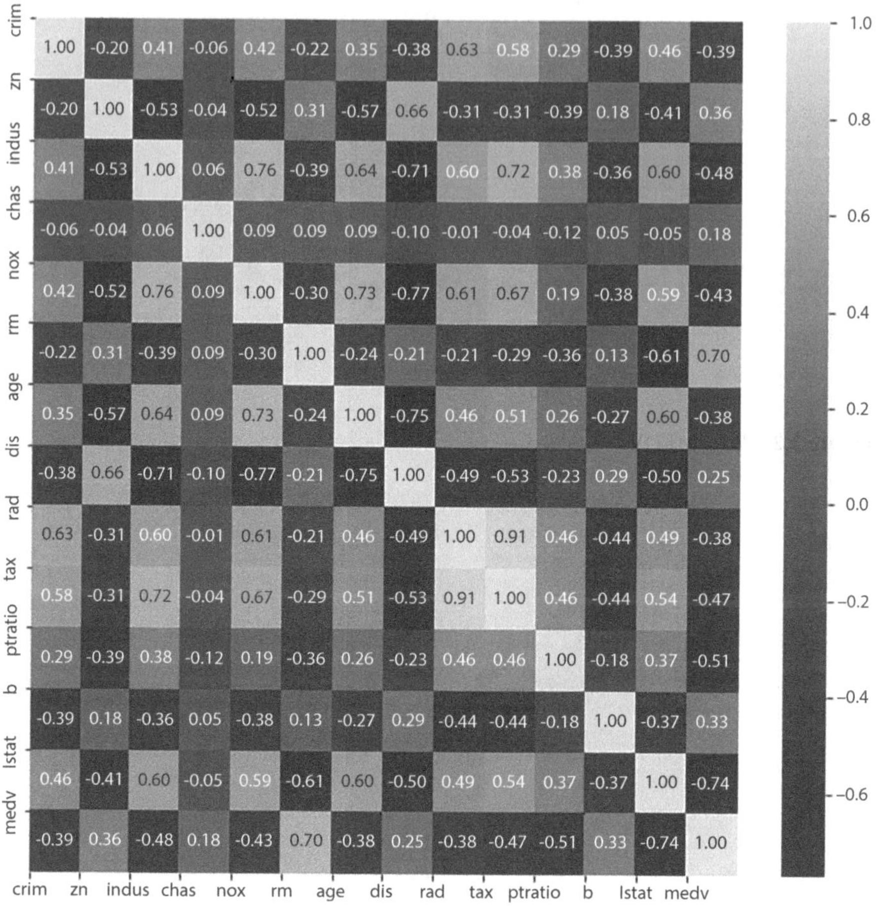

Figure 2.22 Heatmap.

```
x_train, x_test, y_train, y_test = train_test_split(X, Y, test_size=.30,
random_state=10)
# model instantiation
linear_model = LinearRegression()
# fit the model with input and corresponding output data
linear_model.fit(x_train, y_train)
y_pred = linear_model.predict(x_test)
print("Predicted House Prices")
print(y_pred)
```

```
Predicted House Prices
[31.4243217  31.96785487 30.93785448 22.34313349 18.83846235 16.20617519
 35.92908162 14.74157477 25.07700756 37.13230282 21.47652971 30.92661826
 28.07823424 34.02599249 33.7778476  40.63701192 24.25899783 23.43019291
 25.547906   21.34469147 32.65467539 17.80506124 25.46149722 25.0207691
 32.51742137 20.51357936 19.47165255 16.87107974 38.44316206  0.3888111
 32.39559257 32.15518102 26.05305015 23.82049084 20.56494632 19.66990981
  3.53212643 35.21058387 27.03280773 27.67994129 34.36642896 29.82003002
 18.31717228 31.55109654 17.93465111 28.4618882  19.39950216 21.60782793
 38.10391926 16.45101411 24.51003632 19.57072199 24.53359986 34.34589029
 26.74381857 34.86340026 21.02859444 19.77400901 18.68461884 24.64911818
 19.89923131 23.52871967 39.56690084 42.81848202 30.38116596 16.96889789
 23.83550068  3.18952008 31.52469303 28.66536677 18.40003941 27.1252398
 19.56625808 25.28579071 25.04642543 10.30562974 38.97693514  8.17901104
 18.51160158 30.32060192 22.85826904 20.98739809 20.03256896 28.70004491
 30.83986409 28.22575374 26.26715448 31.56221764 22.16821072 -6.40058282
 21.56768106 19.91022274 24.98025707 23.46247865 19.27692546 18.74766072
 27.39474857 22.21870161 26.82617865 23.38366235 23.91947259 19.26003338
 21.09721048 10.97767848 13.82483416 20.72378817 23.53977916 13.92807616
 28.87070782 15.75931299 15.18847286 22.31904753 26.57865025 28.92895886
 24.2980433  18.18585276 16.29471281 17.40018499 15.5361744  21.171008
 33.71665888 30.02422884 21.13474092 13.93326869 16.19662736 29.30890971
 13.1822158  22.06473155 24.33753588 31.89978763 33.44295992  5.90754764
 35.2142013  24.1925307  17.45166483 24.25550318 28.41864261 34.52462088
  6.2544935   2.12089952 28.40739835 12.49858357 18.22016397 19.69247114
  5.52373677 14.45613717 37.18224665 25.78695839 23.32233462 26.36634945
 11.47817153 20.44901468]
```

```
predicted_dataframe = pd.DataFrame(
  {"Predicted_Prices": predicted_house_price, "Actual_Prices": y_test})
print("Predicted and Actual Price Data frame")
print(predicted_dataframe)
```

```
Predicted and Actual Price Data frame
     Predicted_Prices  Actual_Prices
305         31.424322           28.4
193         31.967855           31.1
65          30.937854           23.5
349         22.343133           26.6
151         18.838462           19.6
..                ...            ...
56          25.786958           24.7
37          23.322335           21.0
66          26.366349           19.4
427         11.478172           10.9
12          20.449015           21.7

[152 rows x 2 columns]
```

```
plt.scatter(y_test, predicted_house_price, color='black')
plt.show()
```

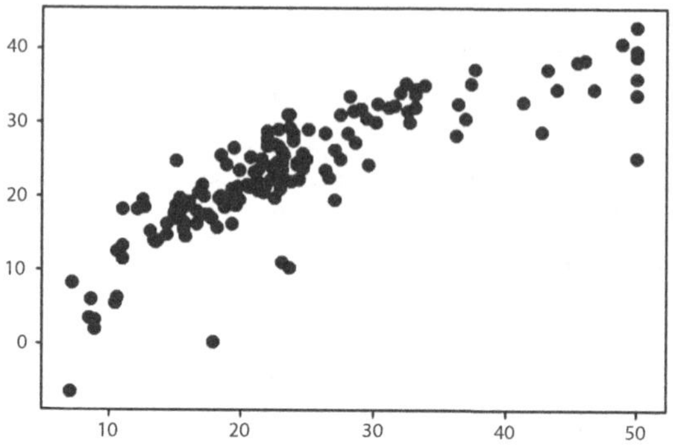

Figure 2.23 Y_test vs. Predicted_House_price.

```
# Get Mean Squared Error (MSE)
from sklearn.metrics import mean_squared_error
mse = mean_squared_error(y_test, y_pred)
# Get Mean Absolute Error (MAE)
from sklearn.metrics import mean_absolute_error
mae = mean_absolute_error(y_test, y_pred)
err = "MSE:" + str(round(mse, 2)) + "," + " MAE:" + str(round(mae, 2))
print("Linear Regression Model Performance:-- ", err)
Linear Regression Model Performance:--  MSE: 29.33, MAE: 3.71
fig = plt.figure(figsize=(12,9))
ax = sns.regplot(y_test, y_pred, marker = 'o', color = 'green')
ax.set_title('Linear Regrassion', fontsize=20)
ax.set_xlabel('MEDV or Price', fontsize=20)
ax.set_ylabel('Predicted Prices', fontsize=20)
```

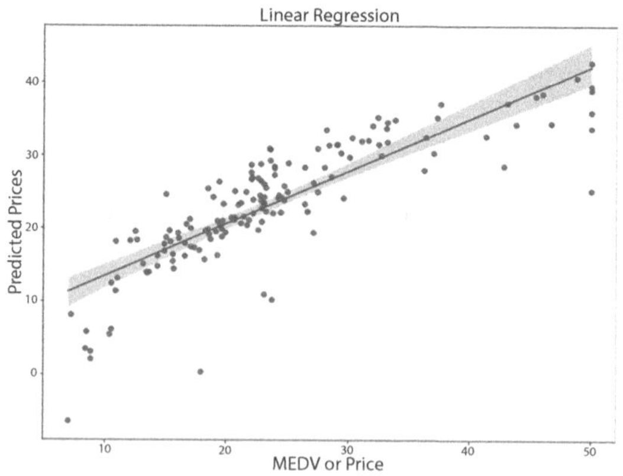

Figure 2.24 Predicted prices vs. MEDV or price.

2.8.6 Gold Price Prediction

In the past, gold had been used as currency in several countries, including the USA. To guarantee the repayment of foreign debts and to limit inflation, which has the effect of revealing a country's financial strength, precious metals like gold are currently held by the central banks of all nations.

Investors can choose the best time to buy (or sell) gold by predicting the rise and decrease in the daily gold rates.

In this project, we would use the most complete set of features to estimate gold rates. We would also use a variety of ML methods and compare the outcomes. We also pinpoint the characteristics that have a significant impact on gold rates.

There are 80 total columns and 1,718 total rows in the data. Data were gathered for attributes like the price of oil, the Standard and Poor's 500 index, the Dow Jones Index US Bond rates (10 years), the exchange rate between the euro and the dollar, the price of precious metals like silver and platinum as well as other metals like palladium and rhodium, the price of the US Dollar Index, and the Eldorado Gold Corporation and Gold Miners ETF.

```
import numpy as np
import pandas as pd
import matplotlib.pyplot as plt
import seaborn as sns
from sklearn.model_selection import train_test_split
from sklearn.ensemble import RandomForestRegressor
from sklearn import metrics
# loading the csv data to a Pandas DataFrame
gold_data = pd.read_csv('gld_price_data.csv')
# print first 5 rows in the dataframe
gold_data.head()
```

	Date	SPX	GLD	USO	SLV	EUR/USD
0	1/2/2008	1447.160034	84.860001	78.470001	15.180	1.471692
1	1/3/2008	1447.160034	85.570000	78.370003	15.285	1.474491
2	1/4/2008	1411.630005	85.129997	77.309998	15.167	1.475492
3	1/7/2008	1416.180054	84.769997	75.500000	15.053	1.468299
4	1/8/2008	1390.189941	86.779999	76.059998	15.590	1.557099

Figure 2.25 Data sample.

```
# print last 5 rows of the dataframe
gold_data.tail()
```

	Date	SPX	GLD	USO	SLV	EUR/USD
2285	5/8/2018	2671.919922	124.589996	14.0600	15.5100	1.186789
2286	5/9/2018	2697.790039	124.330002	14.3700	15.5300	1.184722
2287	5/10/2018	2723.070068	125.180000	14.4100	15.7400	1.191753
2288	5/14/2018	2730.129883	124.489998	14.3800	15.5600	1.193118
2289	5/16/2018	2725.780029	122.543800	14.4058	15.4542	1.182033

Figure 2.26 Data sample.

```
# number of rows and columns
gold_data.shape
(2290, 6)

# getting some basic informations about the data
gold_data.info()
<class 'pandas.core.frame.DataFrame'>
RangeIndex: 2290 entries, 0 to 2289
Data columns (total 6 columns):
 #   Column    Non-Null Count  Dtype
---  ------    --------------  -----
 0   Date      2290 non-null   object
 1   SPX       2290 non-null   float64
 2   GLD       2290 non-null   float64
 3   USO       2290 non-null   float64
 4   SLV       2290 non-null   float64
 5   EUR/USD   2290 non-null   float64
dtypes: float64(5), object(1)
memory usage: 107.5+ KB
```

```
# checking the number of missing values
gold_data.isnull().sum()
```

```
Date       0
SPX        0
GLD        0
USO        0
SLV        0
EUR/USD    0
dtype: int64
```

```
# getting the statistical measures of the data
gold_data.describe()
correlation = gold_data.corr()
```

	SPX	GLD	USO	SLV	EUR/USD
count	2290.000000	2290.000000	2290.000000	2290.000000	2290.000000
mean	1654.315776	122.732875	31.842221	20.084997	1.283653
std	519.111540	23.283346	19.523517	7.092566	0.131547
min	676.530029	70.000000	7.960000	8.850000	1.039047
25%	1239.874969	109.725000	14.380000	15.570000	1.171313
50%	1551.434998	120.580002	33.869999	17.268500	1.303297
75%	2073.010070	132.840004	37.827501	22.882500	1.369971
max	2872.870117	184.589996	117.480003	47.259998	1.598798

Figure 2.27 Data sample describe.

```
# constructing a heatmap to understand the correlation
plt.figure(figsize = (8,8))
sns.heatmap(correlation, cbar=True, square=True, fmt='.1f',annot=True,
annot_kws={'size':8}, cmap='Blues')
```

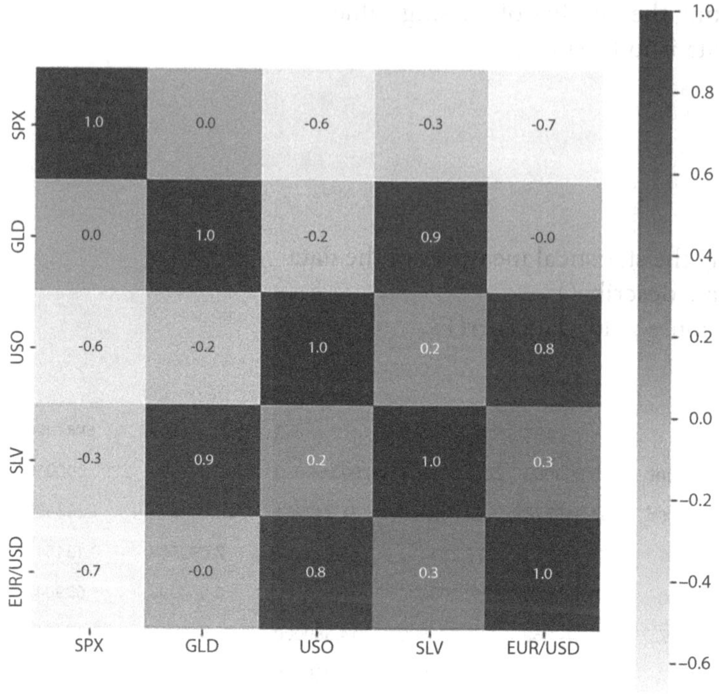

Figure 2.28 Heatmap.

```
# correlation values of GLD
print(correlation['GLD'])
```

```
              SPX        USO      SLV   EUR/USD
0     1447.160034  78.470001  15.1800  1.471692
1     1447.160034  78.370003  15.2850  1.474491
2     1411.630005  77.309998  15.1670  1.475492
3     1416.180054  75.500000  15.0530  1.468299
4     1390.189941  76.059998  15.5900  1.557099
...           ...        ...      ...       ...
2285  2671.919922  14.060000  15.5100  1.186789
2286  2697.790039  14.370000  15.5300  1.184722
2287  2723.070068  14.410000  15.7400  1.191753
2288  2730.129883  14.380000  15.5600  1.193118
2289  2725.780029  14.405800  15.4542  1.182033

[2290 rows x 4 columns]
```

```
# checking the distribution of the GLD Price
sns.distplot(gold_data['GLD'],color='green')
```

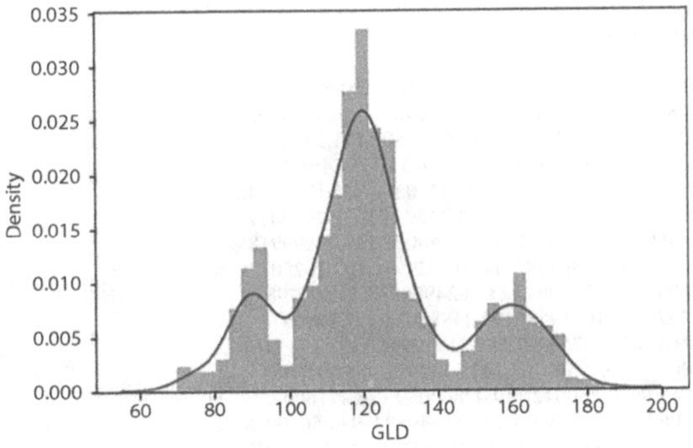

Figure 2.29 Density vs. GLD.

```
X = gold_data.drop(['Date','GLD'],axis=1)
Y = gold_data['GLD']
print(X)
0         84.860001
1         85.570000
2         85.129997
3         84.769997
4         86.779999
            ...
2285     124.589996
2286     124.330002
2287     125.180000
2288     124.489998
2289     122.543800
Name: GLD, Length: 2290, dtype: float64
```

```
print(Y)
RandomForestRegressor()
```

```
X_train, X_test, Y_train, Y_test = train_test_split(X, Y, test_size = 0.2,
random_state=2)
regressor = RandomForestRegressor(n_estimators=100)
# training the model
regressor.fit(X_train,Y_train)
```

```
# prediction on Test Data
test_data_prediction = regressor.predict(X_test)
print(test_data_prediction)
[168.71009953  81.9098999  116.00680049 127.69870074 120.63270109
 154.63659801 150.27749887 126.10790039 117.43529871 126.15300031
 116.54550108 171.66650115 141.55719881 168.00219803 115.29600001
 117.80700027 135.6052037  169.88510107 158.91190311 157.49139984
 155.15890037 125.52840038 175.98959972 157.44710312 125.17700046
  93.99629988  77.85340012 120.43670012 119.08249944 167.59910017
  88.06850031 125.18260002  91.29270058 117.69660021 121.20499918
 135.96190102 115.75610078 115.16670069 147.10929931 107.16280126
 104.36190232  87.27389786 126.52980047 117.82490003 153.19979876
 119.57439982 108.3436997  108.10959881  93.24880078 127.2299975
  75.39730013 113.57429916 121.35769997 111.31959906 119.01739913
 120.89749919 159.05000039 170.17500107 147.3132971   85.80169835
  94.43630035  86.7196986   90.42850042 119.02860039 126.47760073
 127.54460019 169.01680062 122.32349931 117.51549884  98.65610028
 168.1169007  142.96859853 132.09460233 121.27400221 120.70959957
 119.92270076 114.53730158 118.53610043 107.13800101 127.88510092
 114.01939989 107.55300005 116.9565006  119.59509894  88.91770059
  88.177699   146.76540234 127.19940003 113.46100005 110.10409842
 108.28989885  77.91139892 170.25210199 114.04489919 121.68629917
 127.8506019  154.903699    91.74579943 136.85990144 158.09400383
 125.36870064 125.07460022 130.62600148 114.95240134 119.94989962
  92.06519995 110.09519914 167.63449925 156.96679931 114.17979957
 106.78900156  79.66509986 113.25460041 125.87650074 107.19299947
 119.50160122 156.28070285 160.06199914 120.15409992 134.34100299
 100.96360014 117.58919797 119.23799969 112.93370076 102.74749895
 160.28869774  99.33140039 147.85999944 125.44940117 168.81109912
 125.52899919 127.41369753 127.4408024  113.74169926 112.65160074
 123.78639918 102.22059903  88.94489991 124.7370995  101.23239929
 107.15709898 113.53140046 117.54030072  99.48129973 121.70860053
 163.59569906  87.47529874 106.8878999  117.15990103 127.61940158
 124.02540072  80.57699907 120.13170089 158.62549786  87.93359983
 110.33969943 118.85759924 172.12689859 102.92659909 105.49830054
 122.43250037 158.72289759  87.72019836  93.44400079 112.7733003
 177.57809895 114.58959975 119.1766997   95.01530092 125.70420033
 166.03789997 114.90370104 116.84050145  88.24709909 149.16050074
 120.43299966  89.48190002 112.1406003  117.60760079 118.97930109
  88.13489939  94.1777999  116.85070024 118.41300162 120.60240044
 126.63809877 121.94639969 150.83439986 166.37450043 118.63139992
 120.37040137 151.21140059 118.4005991  173.01649891 105.38169945
 105.00370108 150.17140103 113.69690083 124.8311012  147.76649983
 119.71030125 115.34260067 112.6170001  113.42540202 142.06170237
 117.88019747 102.91770043 115.90440129 103.70960185  99.00910068
 117.24220073  90.55920013  91.49310031 153.65839882 102.74399997
 155.11460106 114.42270168 138.99750151  90.15149844 115.52749965
 114.41099967 122.78430045 121.79770022 165.29270049  92.89849916
 135.82960097 121.4459991  120.84650052 104.79080029 143.12270285
 121.3620995  116.58840048 113.47510061 127.34009684 122.9086996
```

```
125.80509987 121.25620026 86.85039906 132.2403022 143.80400223
92.64649989 159.27230021 158.04970267 126.11159924 165.2836995
108.90479956 109.20460064 103.71229811 94.44920021 127.86320293
107.18570017 162.03780007 121.65560028 132.11320104 130.71980092
160.20510033 90.07839884 175.25450188 128.44130034 126.71899868
86.66909934 124.55419963 150.3962971 89.53200022 106.93160001
109.12109981 84.92389864 136.4387997 155.00560314 139.29080371
73.67600057 152.52270064 125.59689918 126.73560001 127.46849907
108.88189943 156.3731008 114.54620101 117.01270166 125.2510993
154.12870112 121.42449988 156.39789893 92.88440056 125.51430115
125.99080064 87.94810076 91.97499922 126.41339921 128.12460391
113.22730077 117.80449765 121.13420002 127.40709748 119.89620092
136.5890009 93.85399918 119.66760051 113.2696011 94.33769953
108.53739986 87.31209915 108.98719936 89.69310008 92.44350024
131.77270281 162.24339992 89.37100001 119.72300072 133.38070207
123.78019985 128.21100171 101.9371983 89.26669865 131.8434999
119.91510009 108.6650997 167.10330038 115.2057005 86.5088993
118.93320076 90.94039937 161.94300085 116.26310039 121.7918001
160.5410985 120.07659927 112.78349961 108.43689866 126.67419983
76.03380026 103.09549992 127.78990286 121.9411994 92.62949982
131.64250034 118.12390153 116.0082998 154.57020272 159.07380084
110.05499936 154.44389813 119.28880087 160.61690016 118.44830026
158.68359891 115.10439942 116.83730023 148.3421995 114.71680076
125.65359885 165.81119874 117.82550026 125.28499924 153.13970369
153.58300226 132.10040077 114.66790046 121.24860212 125.00720069
89.7018007 123.19740035 154.800702 111.67110027 106.68249986
161.91600149 118.60800019 165.60600004 134.18410055 114.92079987
152.93439797 168.52800002 114.61450002 114.09730119 157.25219867
85.38419871 127.12690031 127.90770076 128.74869985 124.00430064
123.7362006 90.60590105 153.23680034 97.13469981 138.13449986
88.9400992 107.5703 115.04730068 112.79060082 124.04639928
91.44219875 125.41220093 162.29369884 119.85569892 165.2065007
126.52459841 112.50900011 127.54709941 95.29389881 91.02980002
103.19439915 120.87009994 83.17729964 126.3318999 159.90990398
117.28580088 118.33339988 119.8991999 122.94649965 120.20000136
121.60400014 118.39870031 107.02149953 148.28969992 125.88829884
115.76010071 74.06560021 127.86490087 154.58720055 122.23380038
125.65730094 88.69930027 103.07339868 124.3730007 120.3034002
73.32430101 151.42799985 121.23090044 104.57520003 86.43439805
115.00609909 172.25019794 119.63670045 160.3621985 113.28249969
120.92590011 118.79290114 96.13899991 118.88750032 125.79039991
118.60759981 96.08550048 154.11570154 122.13910013 147.99519945
159.54010259 114.04220007 122.62119928 149.89129843 127.23910032
165.92660145 135.68010012 120.32339951 167.54079868 108.24199951
121.60089838 139.04870108 106.56109915]
```

```
# R squared error
error_score = metrics.r2_score(Y_test, test_data_prediction)
print("R squared error : ", error_score)
```

```
 R squared error :  0.9887026768255479
```

```
Y_test = list(Y_test)
plt.plot(Y_test, color='blue', label = 'Actual Value')
plt.plot(test_data_prediction, color='green', label='Predicted Value')
plt.title('Actual Price vs Predicted Price')
plt.xlabel('Number of values')
plt.ylabel('GLD Price')
plt.legend()
plt.show()
```

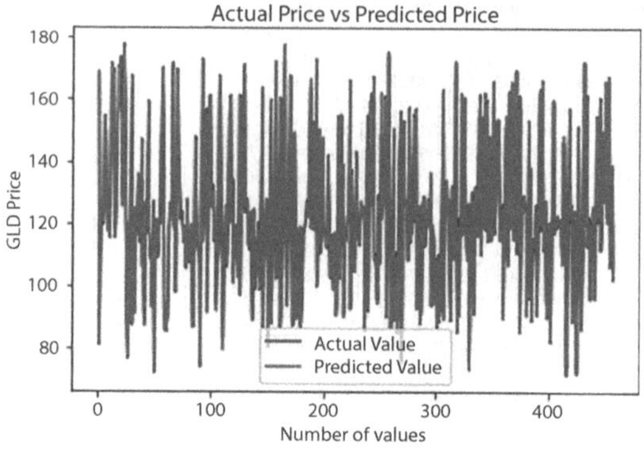

Figure 2.30 Actual price vs. predicted price.

Bibliography

https://www.kdnuggets.com/2021/10/machine-learning-model-development-operations-principles-practice.html

https://www.jeremyjordan.me/evaluating-a-machine-learning-model/

https://machinelearningmastery.com/regression-metrics-for-machine-learning/

https://towardsdatascience.com/what-are-rmse-and-mae-e405ce230383

https://analyticsindiamag.com/step-by-step-building-block-for-machine-learning-models/

https://towardsdatascience.com/3-ways-to-deploy-machine-learning-models-in-production-cdba15b00e

3

Multilayer Perceptron (in Neural Networks)

Multilayer Perceptron (MLP) is simplest type of artificial neural network (ANN). It is a combination of multiple perceptron models. It is a feed-forward class of ANN. Perceptrons are inspired by the human brain and try to simulate its functionality to solve problems. In MLP, these perceptrons are highly interconnected and parallel in nature. This parallelization is helpful in faster computation. Colloquially, MLPs are called as "Vanilla" neural networks, when they have only one hidden layer.

An MLP consists of at least three layers of perceptrons, i.e., an input layer, a hidden layer, and an output layer, which are fully connected. MLPs have the same number of input and output layers but may have multiple hidden layers in between as mentioned in the Figure 3.1. The number of layers is known as the depth, and the number of units in a layer is known as the width [1]. Along with the input, output, and hidden layers, MLPs also have components like weights and biases and activation function. Excluding input layer perceptrons, each perceptron uses a non-linear activation function. Each perceptron in one layer connects with a specific weight to every perceptron in the next layer.

If an MLP has two or more hidden layer, then it is called a deep neural network. Mathematically, the computation of output for the MLP as shown in Figure 3.1 is shown through Equation 3.1 [2, 3]:

$$\hat{y} = \varnothing\left(\sum_{i=1}^{n} w_i \,.\, x_i + b\right) = \varnothing(W^T X + b) \tag{3.1}$$

where $\hat{y}$ is the output of MLP is a vector of shape (m, 1), W is weight matrix of shape (m, n), m is the number of nodes, n is the number of input values, X is the input vector of shape (n, 1), and b is the bias vector of the shape (m, 1).

Kolla Bhanu Prakash. Machine Learning for Industrial Applications, (51–62) © 2024 Scrivener Publishing LLC

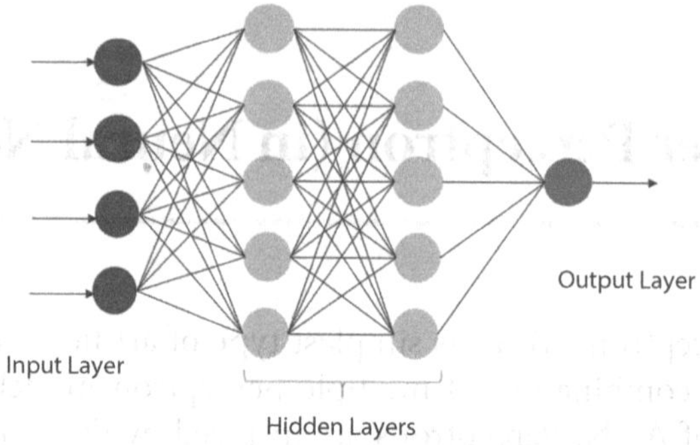

Figure 3.1 Multilayer perceptron with one input, one output, and two hidden layers.

The forward propagation phase [4] involves the chaining of linear function $\lambda()$, $\sigma()$ activation function, in particular, a sigmoid function, τ a threshold function().

The output $\hat{y}$ for the MLP shown in Figure 3.2 is shown in Equation 3.2:

$$\hat{y} = \tau\left(\sigma^{(2)}\left(\lambda^{(2)}\left(\sigma^{(1)}\left(\lambda^{(1)}\left(x_n, w_{mn}\right)\right)\right)\right)\right) \tag{3.2}$$

The output $\hat{y}$ for the MLP shown in Figure 3.3, with another hidden layer, is shown in Equation 3.3:

$$\hat{y} = \tau\left(\sigma^{(3)}\left(\lambda^{(3)}\left(\sigma^{(2)}\left(\lambda^{(2)}\left(\sigma^{(1)}\left(\lambda^{(1)}\left(x_n, w_{mn}\right)\right)\right)\right)\right)\right)\right) \tag{3.3}$$

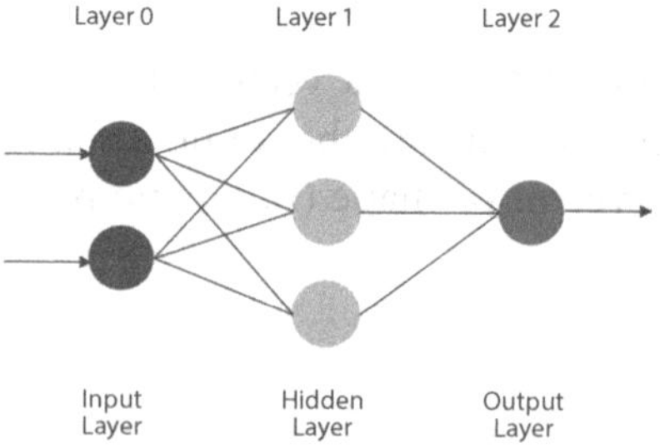

Figure 3.2 MLP with one hidden layer.

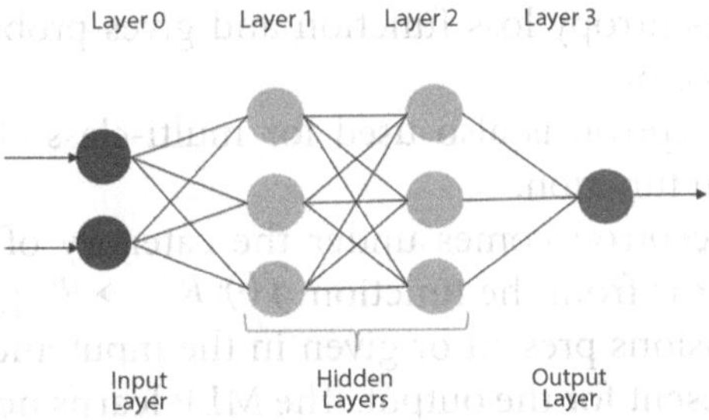

Figure 3.3 MLP with two hidden layers.

```
from sklearn.neural_network import MLPClassifier
X = [[0, 0], [1, 0], [0,1],[1,1]]
y = [0, 1, 0, 1]
# create mutli-layer perceptron classifier
clf = MLPClassifier(solver='lbfgs', alpha=1e-5,hidden_layer_sizes=(5, 3),
random_state=1)
# train
clf.fit(X, y)
# make predictions
print( clf.predict([[1, 2]]) )
print( clf.predict([[-1, 2]]) )
print( clf.predict([[1, -2]]) )
print( clf.predict([[-1, -2]]) )
print( clf.predict([[1.5, -1.2]]) )
```

Output:

```
[1]
[0]
[1]
[0]
[1]
```

The MLP classifier uses an MLP algorithm that is trained based on backpropagation.

The two arrays used by MLP during training are array X of size (n samples, n features), which contains the training samples represented as floating point feature vectors, and array Y of size (n samples), which contains the target values (class labels) for the training data.

Multilayer perceptron in training uses gradient descent in which the backpropagation is used to calculate the gradients. In classification, it reduces the cross-entropy loss function and gives probabilities estimates in the form of a vector.

Multilayer perceptron is also used for multi-class classification using softmax activation function.

Multilayer perceptron comes under the category of supervised algorithms, which learns from the function $f(\cdot): R^m \rightarrow R^o$ given that m is the number of dimensions present or given in the input and o is the number of dimensions present for the output. The MLP learns non-linear function approximator, which can be used for both regression and classification. The difference is that the normal logistic regression has a one or more than one non-linear layers (hidden layers), whereas the logistic regression uses a logistic function, and the MLP uses a step function.

The series of x1, x2,..., xn are the neurons that are on the outermost left region; the layer is called as input layer; from each layer, the value passes by undergoing a weighted linear summation. Finally, in the last layer that is also known as the output layer, the values from the previous layers will be taken to the output layer and transformed into output values.

Figure 3.4 refers to multilayer perceptron representation.

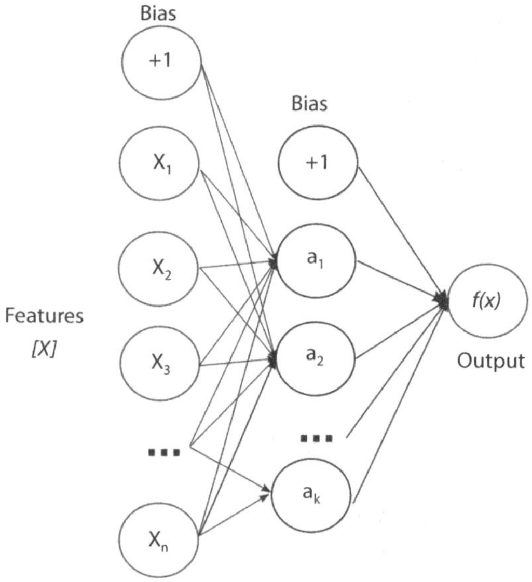

Figure 3.4 Multilayer perceptron representation.

Multilayer perceptron (MLP) has several advantages, which are the capacity to learn non-linear models and also used for online learning by utilizing various parameters like using partial_fit.

With the exception of the fact that we must repeat it twice, once for each group of neurons, and that we must do it layer by layer because, otherwise, the input values to the second layer would not exist, it is essentially identical to the perceptron.

There are several disadvantages of MLP. When there are several local minimums, the loss function for MLP with hidden layers is non-convex. As a result, the validity of different random weight initializations might vary. A variety of hyperparameters, including the quantity of hidden neurons, layers, and iterations, must be tuned for MLP.

The difference between MLP and perceptron is that MLP has at least one hidden layer and MLP can learn both linear and non-linear, but the perceptron can only learn linear functions.

The MLP is modeled after batch algorithm. It computes the average sum-of-squares error, and it is used to update the weights so there is only one set of updates for the epoch. The error gradient is estimated more precisely by the batch technique.

3.1 Multilayer Perceptron for Digit Classification

3.1.1 Implementation of MLP using TensorFlow for Classifying Image Data

The input to the neural network will be grayscale of dimensions 28*28, which will be rescaled/normalized by dividing it with 255. In this model, it consists of one hidden layer, and the other two are input and output layers used, and the model is compiled using the optimizer as adam and loss function to be sparse categorical cross-entropy and used metrics like accuracy to get the model performance and trained the model with 10 epochs, which can be any number but be careful of over fitting, which is a result of over training of the model on the given data and validation and training data split in the ratio 20–80 to be able to get the data in order to test the model.

Then, we are evaluating the model using the in-built method present in the TensorFlow framework [5].

```
# importing modules
import tensorflow as tf
import numpy as np
from tensorflow.keras.models import Sequential
from tensorflow.keras.layers import Flatten
from tensorflow.keras.layers import Dense
from tensorflow.keras.layers import Activation

(x_train, y_train), (x_test, y_test) = tf.keras.datasets.mnist.load_data()
#Loading the Sample MNIST Dataset

# Cast the records into float values
x_train = x_train.astype('float32')
x_test = x_test.astype('float32')

# normalize image pixel values by dividing
# by 255
gray_scale = 255
x_train /= gray_scale
x_test /= gray_scale

#model layers
model = Sequential([

        # reshape 28 row * 28 column data to 28*28 rows
        Flatten(input_shape=(28, 28)),

        # dense layer 1
        Dense(256, activation='sigmoid'),

        # dense layer 2
        Dense(128, activation='sigmoid'),

        # output layer
        Dense(10, activation='sigmoid'),
])
#compiling the model
model.compile(optimizer='adam',
                    loss='sparse_categorical_crossentropy',
                    metrics=['accuracy'])
```

```
#training the model using sample epochs and validation split size
model.fit(x_train, y_train, epochs=10,
                    batch_size=2000,
                    validation_split=0.2)

#evaluating the model
results = model.evaluate(x_test, y_test, verbose = 0)
#printing the test result
```

Output:
test loss, test acc: [0.27210235595703125, 0.9223999977111816]

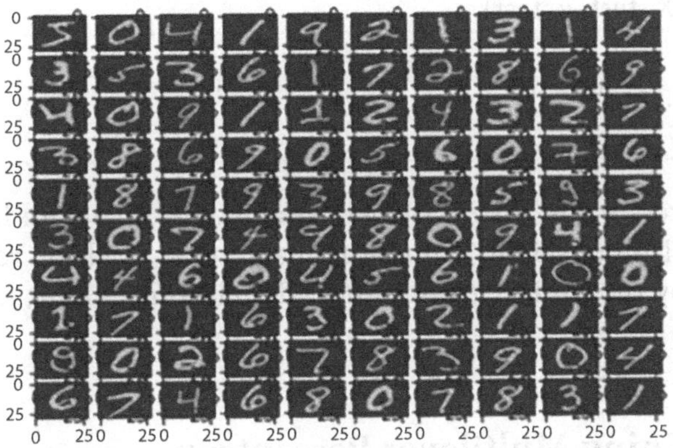

Figure 3.5 MNIST dataset.

The dataset of Modified National Institute of Standards and Technology (MNIST) contains digit images as shown in Figure 3.5 [5].

Multilayer perceptron is also used for regression, in which it trains with no activation function present in the output layer and uses backpropagation technique during training. The backpropagation process can be applied to a network with any number of layers, albeit it becomes increasingly challenging to track which weights are being updated at any given moment. Fortunately, as was already indicated, we almost never require more than two layers. In neural networks, the depth is the total number of layers, whereas the breadth is the number of units in a layer.

Clearly indicating that errors are transmitted backwards over the network is the term "backpropagation of error." It uses a gradient descent technique, in which it uses identity function as activation function, and also the error is calculated by using square error method that results in continuous values.

Multilayer perceptron used in regression [6]:

```
from sklearn.neural_network import MLPRegressor
from sklearn.datasets import make_regression
from sklearn.model_selection import train_test_split
X, y = make_regression(n_samples=200, random_state=1)
X_train, X_test, y_train, y_test = train_test_split(X, y,
                                                    random_state=1)
regr = MLPRegressor(random_state=1, max_iter=500).fit(X_train,
y_train)
regr.predict(X_test[:2])

regr.score(X_test, y_test)
```

```
Output:
array([-0.9..., -7.1...])
0.4...
```

MLP regressor and MLP classifier use alpha parameter for l2 regularization by punishing weights with large magnitudes, and the regularization term used by the MLP regressor and MLP classifier, alpha, helps prevent overfitting.

3.2 Training Multilayer Perceptron

Stochastic gradient descent (SGD), Adam, or L-BFGS are all used in MLP training. With regard to a parameter that requires adaptation, SGD updates parameters using the gradient of the loss function:

$$w \leftarrow w - \eta(\alpha \frac{\partial R(w)}{\partial w} + \frac{\partial Loss}{\partial w})$$

where η is the learning rate, which regulates the parameter space search step size. $Loss$ is applied to the network's loss function.

As a stochastic optimizer, Adam is also almost the same as SGD in the sense that it can alter the number of parameter updates based on adaptive estimations of lower-order moments automatically.

Online and mini-batch learning are supported by training with SGD or Adam.

The second-order partial derivative of a function is represented by the Hessian matrix, and the L-BFGS solver comes close to this. To update parameters, it also makes an approximation of the inverse of the Hessian matrix. L-BFGS is implemented using the Scipy library variant.

If the chosen solver is "L-BFGS," then neither online training nor mini-batch learning is supported.

Multilayer perceptron is also represented as finite acyclic graph.

The weights in the MLP are initially set to small random integers, both positive and negative, according to the MLP algorithm. As we have to take small random integers as we are taking sigmoid activation function, this results in an output of 0 or 1. If the given weights are small in the MLP, then the input is close to 0, which results in output neuron being linear, which results in a linear model. The activity of a neuron is likely to be at, or near, 0 or 1 already if the weights are huge, which suggests that the gradients are modest and that the learning is very slow.

When MLP is used for regression, in that situation, the sigmoid neurons at the output are not particularly helpful. The output neurons can be swapped out with linear nodes that simply add the inputs and use that as their activation. We can also use softmax activation function.

The complete dataset must be passed through the method several times throughout the training of the MLP, with the weights altering as the network makes mistakes throughout each iteration.

We need to train the MLP model until the local minima has been found, but it can lead to overfitting if it is too long; the validation set can be used here to get an approximate of how the network is doing.

3.3 Backpropagation

$$\frac{\partial L}{\partial w_7} = \frac{\partial L}{\partial \breve{y}} * \frac{\partial \breve{y}}{\partial z} * \frac{\partial z}{\partial w_7}$$

Backpropagation is used in the training of the neural network fine-tuning the weights. The gradient of the loss function for a single weight is calculated by the neural network's backpropagation algorithm using the chain rule.

The two types of backpropagation are as follows:

1. Static backpropagation
2. Recurrent backpropagation

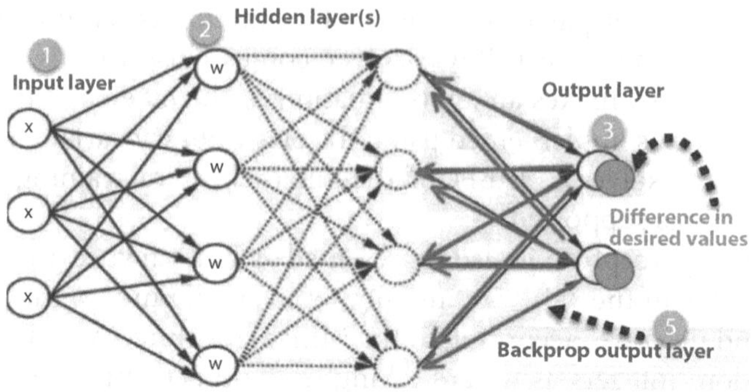

Figure 3.6 Hidden layer representation [7].

Depending on the type of model, we use the backpropagation accordingly

1. Static backpropagation:
   A static input in this network maps to a static output. Static backpropagation will work well for static classification issues like optical character recognition.
2. Recurrent backpropagation:
   Iterative backpropagation is used up until a predetermined threshold is reached. The error is calculated and transmitted backward after the threshold.

Backpropagation has many uses; it requires less memory while computing derivatives than other optimization algorithms, making it memory-efficient. For small- and medium-sized networks in particular, the backpropagation process is quick. It begins to slow down when additional layers and neurons are added and more derivatives are calculated.

This technique is sufficiently general to operate on a variety of network designs, including fully connected networks, generative adversarial networks, and convolutional neural networks.

There is reduced overhead because the backpropagation technique has no tuning parameters. The gradient descent algorithm's only relevant parameters, such as learning rate, are used in the process.

References

1. Grosse, R, Multilayer Perceptrons, https://www.cs.toronto.edu /~mren/teach/ csc411_19s/ lec/ lec10_notes1.pdf.

2. Daniel, H, Mathematical Representation of a Perceptron Layer, https://medium.com/@daniel.hellwig.p/mathematical-representation-of-a-perceptron-layer-with-example-in-tensorflow-754a38833b44.
3. Honkela, A, Multilayer Perceptrons, https://users.ics.aalto.fi/ ahonkela/dippa/node41.html.
4. Pablo, The Multilayer Perceptron - Theory and Implementation of the Backpropagation Algorithm, https://pabloinsente.github.io/the-multilayer-perceptron.
5. Multi-Layer Perceptron Learning in Tensorflow, https://www.geeksforgeeks.org/multi-layer-perceptron-learning-in-tensorflow/.
6. cikit learn https://scikit-learn.org/stable/modules/generated/sklearn.neural_network.MLPRegressor.html#sklearn.neural_network.MLPRegressor
7. Back Propagation in Neural Network, https://www.guru99.com/backpropogation-neural-network.html [last Accessed on 8.11.2022]

4
Kernel Machines

4.1 Different Kernels and Their Applications

In machine learning, kernel machines (KMs) are class pattern analysis algorithms. Their purpose is to find and study the types of relations in datasets. Kernel machines receive their name from the fact that they work in the feature space using kernel functions rather than computing the coordinates of the data in that space. Instead, they compute the inner products between the images of all pairs of data in the feature space. This method is often less computationally demanding than precisely calculating the coordinates. This approach is called the "**kernel trick**" [1]. Kernel functions have been added for sequence data, text, images, and vectors.

The following are the different types of kernel functions [3, 4]:

(i) Polynomial Kernel: The polynomial kernel is a kernel function that represents the similarity of vectors (training samples) in a feature space over polynomials of the original variables, allowing non-linear models to be learned. Support vector machines (SVMs) and other kernelized models are frequently employed with it.

$$k(\mathbf{x}_i, \mathbf{x}_j) = (\mathbf{x}_i \cdot \mathbf{x}_j + 1)^d$$

(ii) Sigmoid Kernel: Its primary application is in neural networks. In a two-layer perceptron model of a neural network, this kernel function is analogous to the activation function for neurons.

$$k(x, y) = \tanh(\alpha x^T y + c)$$

Kolla Bhanu Prakash. Machine Learning for Industrial Applications, (63–72) © 2024 Scrivener Publishing LLC

(iii) Linear Kernel: It is the most basic type of kernel and generally has a one-dimensional shape. It is the finest function when there are multiple qualities. Because most text classification problems can be partitioned linearly, the linear kernel is often utilized. In comparison to linear kernel functions, other functions are slower.

$$K(x, y) = x^T y + c$$

(iv) Radial Basis Function (RBF) Kernel: The RBF kernel, often known as the RBF kernel, is a prominent kernel function that is utilized in a variety of kernelized learning techniques. It is most typically used in SVM classification.

$$K(\mathbf{x}, \mathbf{x}') = \exp\left(-\frac{||\mathbf{x} - \mathbf{x}'||^2}{2\sigma^2}\right)$$

```
from sklearn.pipeline import make_pipeline
from sklearn.preprocessing import StandardScaler
from sklearn.svm import SVC
import numpy as np

#Input and Output
X = np.array([[-2,-2],[-1, -1], [-2, -1], [1, 1], [2, 1], [2, 2]])
y = np.array([1, 1, 2, 2, 1, 2])

#Polynomial Kernel
clf = make_pipeline(StandardScaler(), SVC(gamma='auto',kernel='poly'))
clf.fit(X, y)
print(clf.predict([[-1.5, 1.5]]))

#sigmoid kernel
clf = make_pipeline(StandardScaler(), SVC(gamma='auto',kernel='sigmoid'))
clf.fit(X, y)
print(clf.predict([[-1.5, 1.5]]))
```

```
#linear kernel
clf = make_pipeline(StandardScaler(), SVC(gamma='auto', kernel='linear'))
clf.fit(X, y)
print(clf.predict([[-1.5, 1.5]]))
#Radial Basis Function kernel
clf = make_pipeline(StandardScaler(), SVC(gamma='auto',kernel='rbf'))
clf.fit(X, y)
print(clf.predict([[-1.5, 1.5]]))
```

```
Output:
[2]
[2]
[2]
[2]
```

The kernel technique approach uses high-dimensional feature space while maintaining linear functions.

A kernel function may be used to quickly calculate the pairwise inner products from the original data items. Using a linear classifier and kernel, we can separate data with a non-linear decision boundary.

The traditional Euclidean inner product in kernel methods can be expanded to a flexible nonlinear inner product, represented by a Mercer kernel. Figure 4.1 represents Kernel separation in higher dimensions.

Due to the random distribution of the points, there is no appropriate linear line that should be able to identify the red and green dots, making it

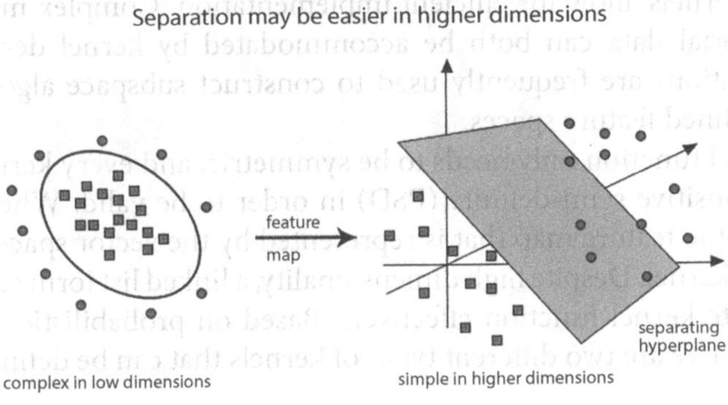

Figure 4.1 Kernel trick [5].

highly challenging to solve this classification problem using a linear classifier. Here, the kernel function is utilized to move the points to higher dimensions, address the issue there, and return the result.

For example, the kernel methods convert a two-dimensional plane that was previously categorized by a one-dimensional hyperplane ("or a straight line") into a three-dimensional area. In this case, the hyperplane, or classifier, is a two-dimensional plane that cuts the area instead of a straight line.

The distance between two patterns, x and y, can be described using the kernel function. The polynomial kernel, Fisher's kernel, diffusion kernel, and the k-spectrum kernel or string kernel are a few of the kernels that have been employed. The pairwise relationship in kernelized learning models just needs to be well stated in order for it to be suitable for both vectorial and nonvectorial data analysis.

Kernel methods are also used in feature extraction. The foundation of kernel methods is the use of kernel-defined feature spaces for a range of learning tasks and problems.

Depending on the structure of the data, kernels are used to attain results.

Kernels are used in structured data to compute even when the number of features is exponential and can compute the products very efficiently using dynamic programming.

Kernels are used for probabilistic models like for smoothly parameterized models to attain efficient result. Fisher's kernel used P-kernels that take into account a family of models, each of which offers one characteristic equal to the likelihood of the example in that model.

In chemometrics, a multidimensional regression approach known as kernel partial least squares is used.

To learn a semantic representation independent of the two views, kernel canonical correlation analysis uses paired datasets. Dual representations in kernels allow for efficient implementation. Complex models and non-vectorial data can both be accommodated by kernel design. Dual representations are frequently used to construct subspace algorithms in kernel-defined feature spaces.

A kernel function only needs to be symmetric, and every kernel matrix must be positive semi-definite (PSD) in order to be valid. When dealing with text, the feature map that is represented by the vector space model is called the kernel. Despite high dimensionality, a linked list form can be used to compute kernel function effectively. Based on probabilistic models of the data, there are two different types of kernels that can be defined, which are known as marginalization kernel and also Fisher's kernels. Fisher's kernels are also used based on string kernels and also hidden Markov models.

In the kernel method approach, the data are inserted into a Hilbert space or Euclidean feature space, and, also, when the data are given as the images, the relations that are drawn from the images by the kernel methods are linear relations.

Embedding is created to enable the efficient "shortcut" known as the kernel to compute the inner products of pictures of two points directly. With the help of kernel methods, it can easily conduct linear regression in spaces with very high dimensions and can also even work with infinite dimensions with the help of Gaussian kernel methods.

The subspace methods are further implemented using kernel, which is to represent the principal eigenvectors and the projection function. Kernel methods are the go-to functions when we are dealing with pattern analysis. In order to allow users to embed previous knowledge into a measure of similarity between two objects, kernels create flexible interfaces to the data [1].

Kernels are implemented in various subspace methods such as kernel principal component analysis, kernel gram Schmidt, and also kernel canonical correlation analysis that is used to identify nonlinear relationships between two blocks of multivariate data, which in turn maximizes the correlation between the data points projected into the data space [7].

The most popular kernel function is the RBF, as it responds in a finite manner throughout the whole x-axis. Various kernels methods are used in SVM such as linear, sigmoidal, RBF, and also the non-linear and the polynomial functions. By using the kernel method, nonlinear SVM successfully increases the dimensionality of the patterns by obtaining a linear decision boundary in a higher-dimensional space.

The kernel functions are also known as the generalized dot products as they calculate the dot product in higher dimension feature space. In terms of a dual representation, many linear models for regression and classification may be recast, and the kernel function appears naturally.

4.2 Some Other Kernel Functions

4.2.1 Gaussian Radial Basis Function (RBF)

When the data are not known, the Gaussian RBF is used. The value of an RBF kernel relies on how far it is from the origin or another location. Figure 4.2 represents the Gaussian kernel visualization.

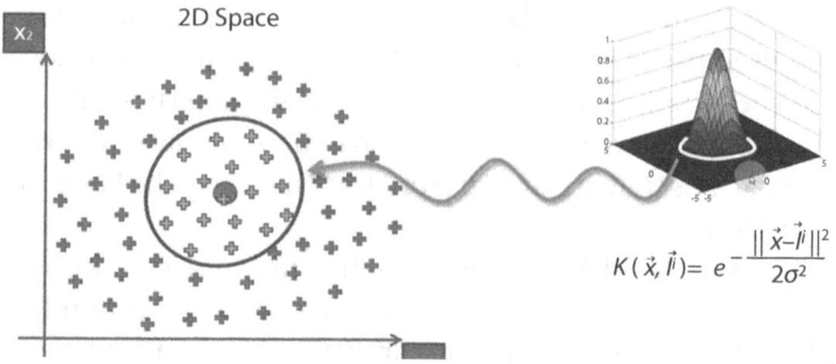

Figure 4.2 Gaussian kernel visualization [7].

$$k(\mathbf{x_i}, \mathbf{x_j}) = \exp(-\gamma \, ||\mathbf{x_i} - \mathbf{x_j}||^2)$$

4.2.2 Laplace RBF Kernel

When the data are not known, Laplace RBF kernel is used. The value of an RBF kernel relies on how far it is from the origin or another location.

$$k(x, y) = \exp\left(-\frac{||x - y||}{\sigma}\right)$$

4.2.3 Hyperbolic Tangent Kernel

When we are dealing with neural networks, we will use hyperbolic tangent kernel. Figure 4.3 gives laplace kernel visualization.

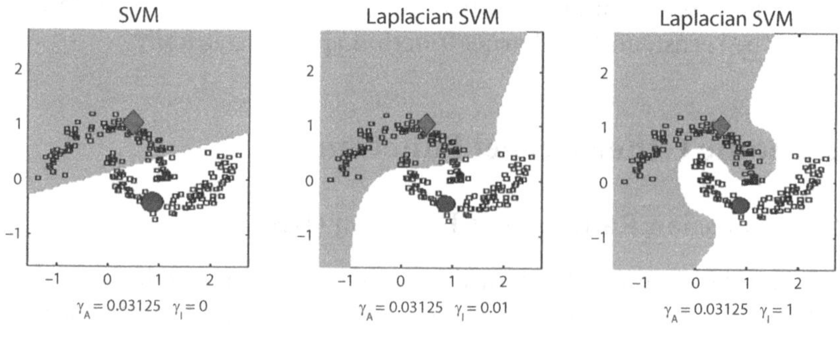

Figure 4.3 Laplace kernel visualization.

$$k(\mathbf{x_i},\mathbf{x_j}) = \tanh(\kappa \mathbf{x_i} \cdot \mathbf{x_j} + c)$$

4.2.4 Bessel Function of the First-Kind Kernel

Bessel function of the first-kind Kernel can be applied to mathematical functions to eliminate the cross term.

$$k(x,y) = \frac{J_{v+1}(\sigma\,||x-y||)}{||x-y||^{-n(v+1)}}$$

4.2.5 ANOVA Radial Basis Kernel

ANOVA radial basis kernel is used when dealing with regression problems [6].

$$k(x,y) = \sum_{k=1}^{n} \exp(-\sigma(x^k - y^k)^2)^d$$

4.2.6 Linear Splines Kernel in One Dimension

When working with huge sparse data vectors, it is helpful. It is frequently employed in text classification. Additionally, the splines kernel does well in regression issues. Figure 4.4 represents ANOVA radial basis kernel.

$$k(x,y) = 1 + xy + xy\,min(x,y) - \frac{x+y}{2} min(x,y)^2 + \frac{1}{3}\min(x,y)^3$$

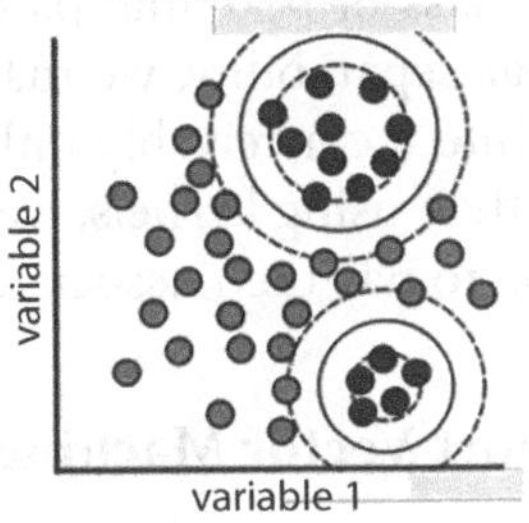

Figure 4.4 ANOVA radial basis kernel.

4.2.7 Exponential Kernel

This is the same as Gaussian kernel but with square removed.

$$K(x,y) = \exp(-\frac{||x-y||}{2\sigma^2})$$

Because adding dimension to these datasets does not always increase separability, the linear kernel is frequently employed on datasets with a lot of characteristics.

The polynomial kernel function is used when image data are present to make the model based on image processing, whereas, when we do not have any idea on the data given, we will use the Gaussian kernel; it is also called as general purpose kernel, and there is also the Gaussian RBF and Laplace RBF that are also the kernel methods that are used when the data are not known. These are types of radial basis kernel methods are frequently used in SVMs.

Other kernel methods are hyperbolic tangent kernel, which is used when we are modeling deep learning models specifically the neural networks; we will also use sigmoid kernel function in the same manner as the hyperbolic tangent kernel and Bessel function of the first kind is used when there are mathematical equations involved.

ANOVA radial basis kernel method that we will use for regression problems and, lastly, the linear splines kernel in one dimension are useful when the data modeled are in the form of large sparse data vectors; the linear splines kernel in one-dimensional kernel method is often used when we are modeling the text data, which comes under the text categorization like grouping of words. The linear splines kernel in one-dimensional kernel method is also similar to the ANOVA radial basis kernel used for modeling many regression problems.

By providing a concept of data similarity, the class of kernel techniques implicitly determines the class of potential patterns. In order to identify a dimension that has linear separability, we may use a kernel to assist us in transforming our data into a conceivably unlimited number of dimensions. We also discovered that, using kernels, we could access these higher dimensions without having to pay their associated costs.

4.2.8 Kernels in Support Vector Machine

Support vector machine is very popular machine learning model that is used for classification and regression as well as detecting the outliers [2].

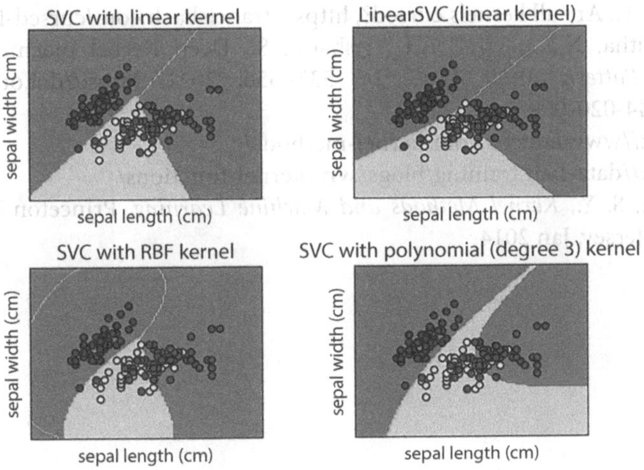

Figure 4.5 SVM classification with different kernels.

Figure 4.5 represents SVM classification with different kernels. Support vector machine uses hyperplane in order to classify the data when the data are present in the data plane by finding the optimal hyperplane to classify the data points.

The SVMs work well when the data are linear, but, when the data are non-linear, the division/classification of the data points becomes difficult as there are more data clusters or points to classify In order to overcome the above situation, we use kernel trick in the SVM to classify the non-linear data points.

The kernel trick used in SVM is nothing, but, if the data are impossible to separate in the current given dimensions, then add additional dimension to the plane at which the data points are present.

Example: If the data given is in two dimensions, which is inseparable, then we will add another dimension and make it into three dimensions, which results in our desired classification, but the dimension adding is based on the distance present between the data points present in the plane.

References

1. Shawe-Taylor, J. and Cristianini, N., *Kernel Methods for Pattern Analysis*, Cambridge University Press, University of Southampton, Nello Cristianini, University of California, Davis, 2004.
2. Lendave, I., How do Kernels Learn Non-Linear Functions, https://analyticsindiamag.com/how-do-kernels-learn-non-linear-functions/.

3. Bach, F., Are all kernels cursed?, https://francisbach.com/cursed-kernels/.
4. Nikhitha, N.K., Afzal, A.L., Asharaf, S., Deep Kernel machines: a survey. *Pattern Anal. Appl.*, 24, 537–556, 2021, https://doi.org/10.1007/s10044-020-00933-1.
5. https://www.educba.com/kernel-methods/
6. https://data-flair.training/blogs/svm-kernel-functions/
7. Kung, S. Y., *Kernel Methods and Machine Learning*, Princeton University, New Jersey, Jan 2014.

5

Linear and Rule-Based Models

5.1 Least Squares Methods

The least square approach involves minimizing the sum of the squares of the offsets (residual component) of the points from the curve to get the best-fitting curve or line of best fit for a group of data points. The trend of outcomes is statistically evaluated throughout the process of determining the relationship between two variables. Regression analysis is the name for this procedure. Curve fitting is a technique for doing regression analysis. The least squares approach of fitting equations approximates the curves to provided raw data.

Depending on whether the residuals in all unknowns are linear, there are two forms of least squares problems: linear or ordinary least squares and nonlinear least squares. The linear least squares problem arises in statistical regression analysis; it has a closed-form solution. When dealing with nonlinear issues, iterative refinement is utilized; the system is approximated by a linear one at each iteration, and the core computation is the same in both cases.

5.2 The Perceptron

In a linear classification algorithm, the perceptron is used to classify data. This implies that it learns a decision boundary in the feature space that divides two classes using a line (known as a hyperplane). As a result, it is best suited to issues in which the classes can be easily separated using a line or linear model, sometimes known as linearly separable problems [1]. It is made up of a single node or neuron that processes a row of data and predicts a class label. The weighted total of the inputs and a bias are used to achieve this (set to 1). The activation is defined as the weighted sum of the model's input [1].

Kolla Bhanu Prakash. Machine Learning for Industrial Applications, (73–86) © 2024 Scrivener Publishing LLC

$$A = W X I + B$$

where A is the activation, W is the weight, I is the input, and B is the bias.

If the activation is above 0.0, then the model will output 1.0; otherwise, it will output 0.0 [1].

5.2.1 Bias

The bias term is a numerical term that may be adjusted and is added to the weighted sum of inputs and weights in a perceptron to improve classification model accuracy. The default input value for the *bias* weight is 1, and the weight value is adjustable.

5.2.2 Perceptron Weighted Sum

Calculating the weighted sum of the perceptron's inputs and weights is the first step in the perceptron classification process. To accomplish this, multiply each input value by the corresponding weight and then add all of the resulting products. Based on their relative importance, this sum provides a suitable representation of the inputs.

5.2.3 Activation Function

Depending on the case, an activation function is applied to the weighted sum of the inputs and weights to limit the perceptron output to a value within a specific range. The perceptron output is given to the activation

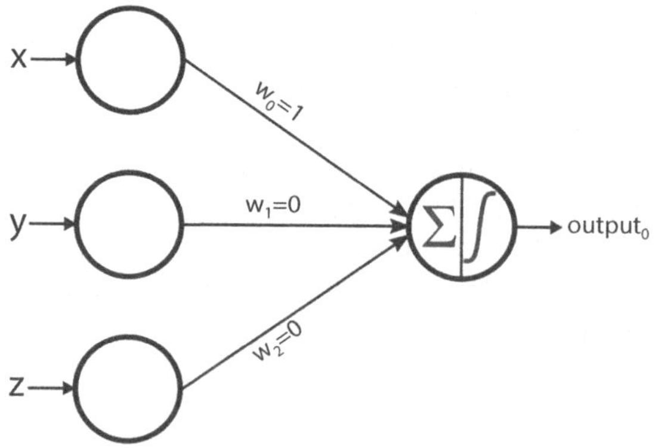

Figure 5.1 Perceptron diagram.

function, and the activation function returns the output value either 1 or −1.

- If weighted sum is greater than 0, then the activation function returns 1.
- If weighted sum is less than 0, then the activation function return −1.

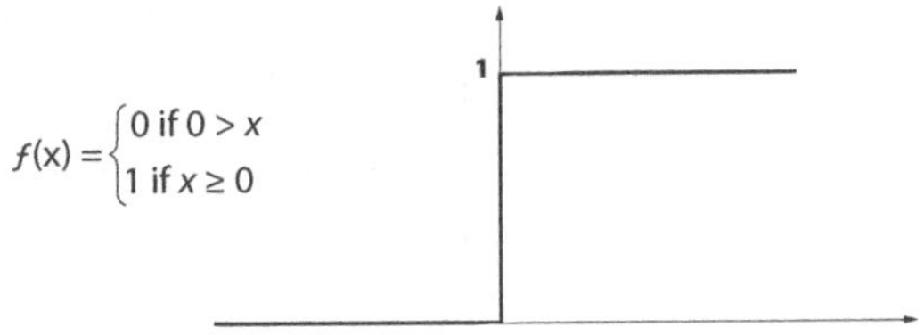

5.2.3.1 *Types of Activation Functions*

There are 10 types of activation functions that we use them regularly. They are as follows [4]:

1. Binary step function
2. Linear function
3. Sigmoid
4. Tanh
5. ReLU
6. Leaky ReLU
7. Parameterized ReLU
8. Exponential linear unit
9. Swish
10. Softmax

5.2.3.1.1 Binary Step Function

When we think of an activation function, the first thing that springs to mind is a threshold-based classifier, which determines whether or not to activate the neuron depending on the result of the linear transformation. The neuron is activated if the input to the activation function exceeds a threshold; otherwise, it is deactivated, which means that its output is not taken into account for the next hidden layer.

$$F(x) = 1, x \geq 0$$
$$= 0, x < 0$$

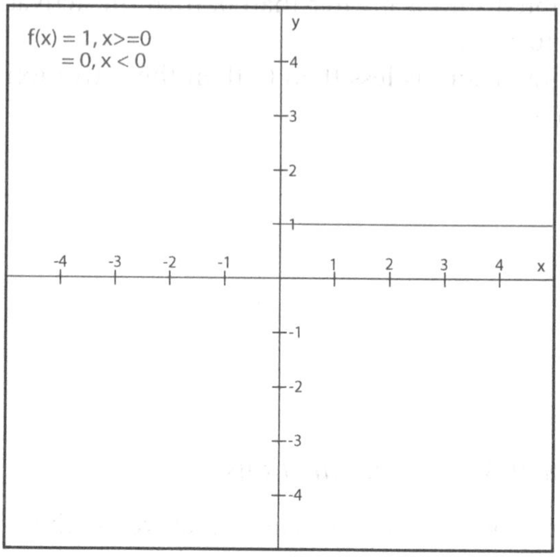

5.2.3.1.2 Linear Function

The step function had an issue; the gradient of the function was zero. This is due to the binary step function not having an x component. We can employ a linear function as opposed to a binary function.

$$F(x) = ax$$

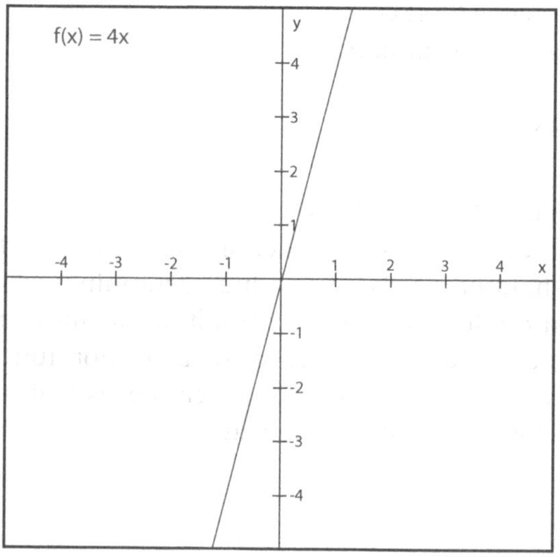

5.2.3.1.3 Sigmoid

The sigmoid function is the following activation function that we will examine. It is one of the most popular non-linear activation functions. The sigmoid function changes values between 0 and 1.

$$F(x) = 1/(1 + e^{\wedge}{-x})$$

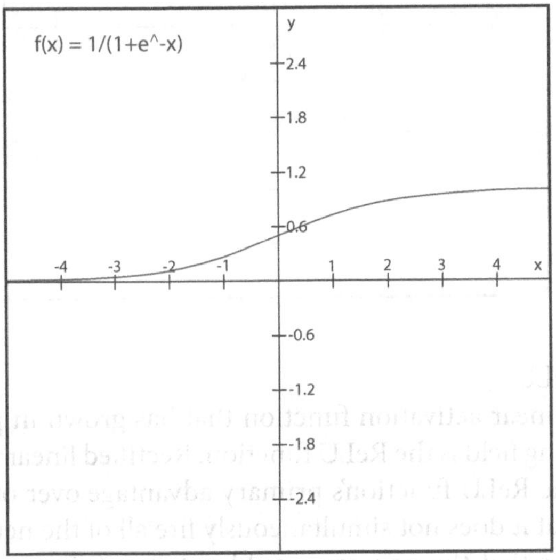

5.2.3.1.4 Tanh

The tanh function and the sigmoid function are quite similar. The symmetry around the origin is the only variation. In this instance, the possible values are between −1 and 1. As a result, the inputs to the following layers would not always have the same sign.

$$Tanh(x) = (2/(1 + e^{\wedge}{-2x})) - 1$$

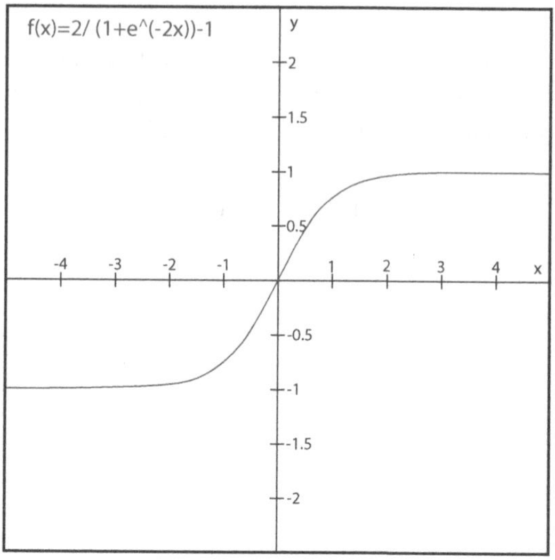

5.2.3.1.5 ReLU

Another non-linear activation function that has grown in prominence in the deep learning field is the ReLU function. Rectified linear unit is referred to as ReLU. The ReLU function's primary advantage over other activation functions is that it does not simultaneously fire all of the neurons.

This signifies that the neurons would not stop firing unless the linear transformation's result is less than 0.

$$F(x) = \max(0, x)$$

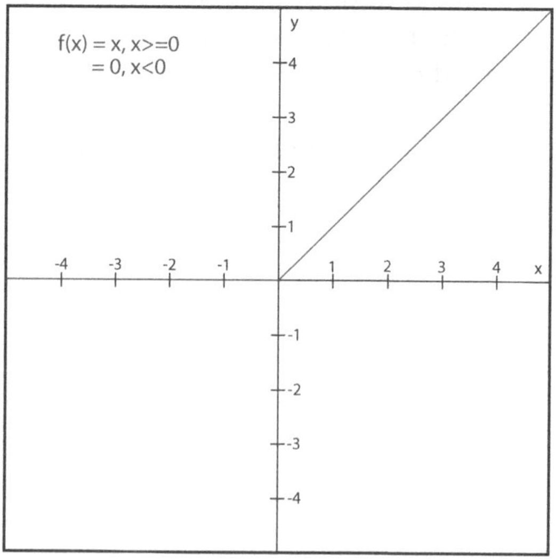

5.2.3.1.6 Leakly ReLU

A better variant of the ReLU function is what the leaky ReLU function is. As we previously observed, the gradient for the ReLU function is 0 for x = 0, which would deactivate the neurons in that area.

To solve this issue, leaky ReLU is defined. We define the ReLU function as a very small linear component of x rather than as 0 for negative values of x.

$$F(x) = 0.01x, x > 0$$

$$= x, x \geq 0$$

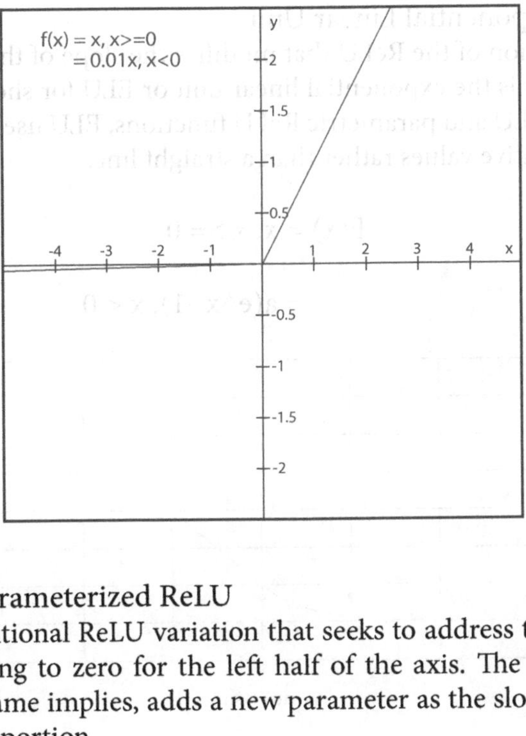

5.2.3.1.7 Parameterized ReLU

This is an additional ReLU variation that seeks to address the issue of the gradient turning to zero for the left half of the axis. The parameterized ReLU, as its name implies, adds a new parameter as the slope of the function's negative portion.

$$F(x) = x, x \geq 0$$

$$= ax, x < 0$$

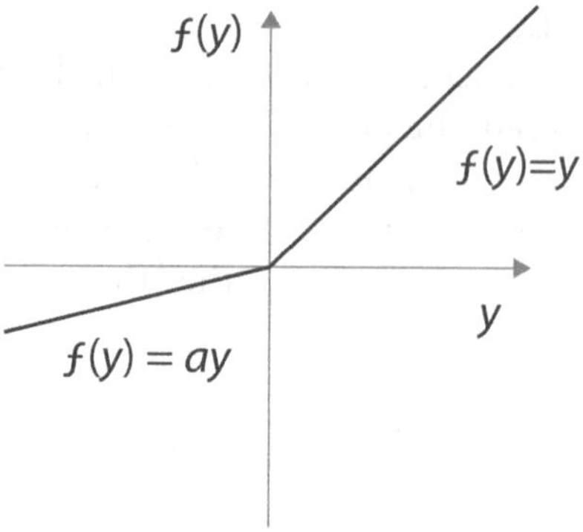

5.2.3.1.8 Exponential Linear Unit

Another variation of the ReLU that modifies the slope of the negative half of the function is the exponential linear unit or ELU for short. In contrast to the leaky ReLU and parametric ReLU functions, ELU uses a log curve to define the negative values rather than a straight line.

$$F(x) = x, x >= 0$$

$$= a(e^x-1), x < 0$$

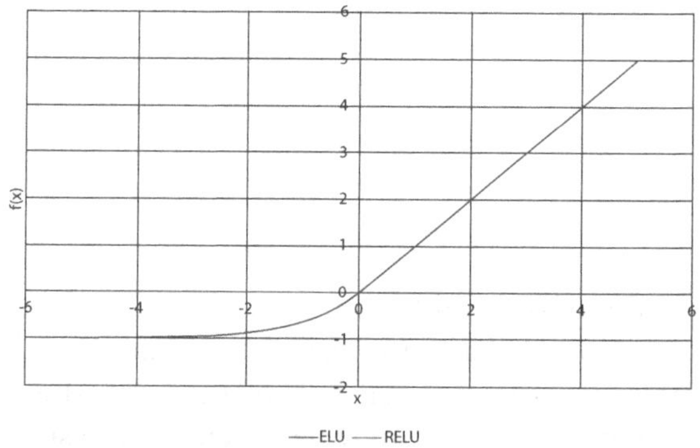

5.2.3.1.9 Swish

Swish is a less well-known activation function that was found by Google researchers. Swish exhibits superior performance than ReLU on deeper models and is as computationally efficient as ReLU. Swish has values ranging from –infinity to infinity.

$$F(x) = x^*\text{sigmoid}(x)$$

$$F(x) = x/(1 - e^\wedge{-x})$$

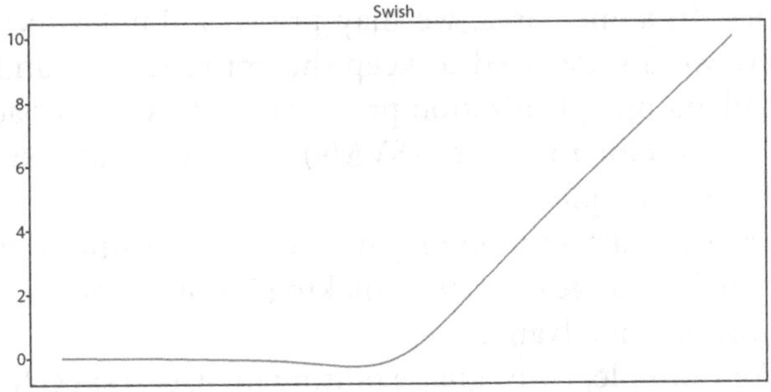

5.2.3.1.10 Softmax

The softmax function is frequently referred to as a concatenation of different sigmoids. We are aware that the sigmoid function returns values in the range of 0 and 1, which can be interpreted as probability that a data point belongs to a specific class. As a result, sigmoid is frequently employed for binary classification issues.

For issues involving multiclass classification, the softmax function can be employed. The chance that a data point belongs to each distinct class is returned by this function.

$$\sigma(\mathbf{z})_j = \frac{e^{z_j}}{\sum_{k=1}^{K} e^{z_k}} \quad \text{for } j = 1, \ldots, K.$$

5.2.4 Perceptron Training

The neural network perceptron is only a method of implementing the hyperplane that the perceptron describes. If you have a sample of data, then you may compute the weight values in advance and then input them into the perceptron to get the output values.

In order to train neural networks, we often employ online learning, where we get individual instances rather than the entire sample. We want the network to update its parameters after each occurrence and gradually adapt over time. Such an approach is interesting for a number of reasons:

1. We do not have to pay to store the training sample in an external memory or the optimization's intermediate outcomes. In some cases, we may prefer a simpler technique where we do not need to keep the entire sample and solve a challenging optimization problem on it. An approach like support vector machines (SVMs) may be rather expensive with huge samples.
2. The sample distribution may not be constant since the problem could change over time, making it impossible to select a training set in advance.

   For example, we might be putting in place a speech recognition system that adjusts to the user.
3. The system could undergo physical modifications. For example, a robotic system's parts could deteriorate or its sensors could stop working properly.

5.2.5 Online Learning

In online learning, we write the error function on individual instances rather than the entire sample. Starting with arbitrary initial weights, we tweak the parameters slightly after each cycle to reduce error while keeping in mind what we have already learnt. If this error function can be differentiated, then gradient descent can be used.

For example, in regression, the error on the single instance pair with index t, (xt , rt), is

$$Et\,(w|\overset{t}{x},\overset{t}{r}) = \tfrac{1}{2}(\overset{t}{r}-\overset{t}{y})^2 = \tfrac{1}{2}[\overset{t}{r}-(w^T\,\overset{t}{x})]^2$$

and for j = 0,...,d, the online update is $\Delta \overset{t}{w} = \eta(\overset{t}{r} - \overset{t}{y})\overset{t}{x}\,j$

where η is the learning factor, which is progressively lowered as convergence approaches. Stochastic gradient descent is the term used for this.

Similar to this, update rules can be constructed for classification issues using logistic discrimination, where updates are made after each pattern rather than after a full run over the training set and instead are made after the updates are added up.

With two classes, for the single instance (xt, rt) where r t i = 1
if xt $\in$ C1 and r t i = 0 if xt $\in$ C2,
the single output is yt = sigmoid(wT xt)
and the cross-entropy is Et (w|xt , r t) = −r t log yt − (1 − r t) log(1 − yt)
We obtain the following online update rule by using gradient descent for j = 0,...,d:
Δwt j = η(r t − yt)xt j
When K > 2 classes, the outputs are for the single instance (xt, rt), where r t I = 1 if xt $\in$ Ci and 0 otherwise.
yt i = exp wT i xt / Σkexp wT k xt
and the cross-entropy is
Et ({wi}i|xt , rt) = −Σi r t i log yt

5.2.6 Perceptron Training Error

A perceptron's classification accuracy for a particular training data sample is indicated by its training error. It simply quantifies "how poorly" the perceptron is performing and aids in identifying which weight changes are necessary to improve classification accuracy [3].

$$\text{training\_error} = \text{actual\_label} - \text{predicted\_label}$$

A perceptron should aim for a training error of zero because this shows how well it is working with a sample of data.

The example of perceptron is shown below [1]:

```
# make a prediction with a perceptron model on the dataset
from sklearn.datasets import make_classification
from sklearn.linear_model import Perceptron
# define dataset
X, y = make_classification(n_samples=100000, n_features=10, n_info
rmative=10, n_redundant=0, random_state=1)
# define model
model = Perceptron()
# fit model
model.fit(X, y)
# define new data
row1 = [1.12777556,-12.64400522,2.23268854,0.89834386,-
11.75498361,2.1243966,-1.08897217, -6.35123476, 5.00005655,-
3.82764385]
row2 = [0.12777556,-12.64400522,0.23268854,0.89834386,-
11.75498361,0.1243966,0.08897217, -1.35123476, 0.00005655,-
3.11114385]
# make a prediction
yhat1 = model.predict([row1])
yhat2 = model.predict([row1])
# summarize prediction
print('Predicted Class: %d' % yhat1)
print('Predicted Class: %d' % yhat2)
```

Example for Perceptron

Output

```
Predicted Class: 1
Predicted Class: 1
```

[1] Jason Brownlee, Perceptron Algorithm for Classification in Python, https://machinelearningmastery.com/perceptron-algorithm-for-classification-in-python/ [last accessed on 31 December 2021].

5.3 Support Vector Machines

Support vector machines (SVMs) are a collection of supervised learning algorithms used for classification and regression analysis utilizing data analysis and pattern recognition. Methods differ depending on the classifier's structure and features. The most well-known SVM is a linear classifier, which predicts the member class of each input between two potential classifications. A better description would be that a SVM constructs a hyperplane or group of hyperplanes to categorize all inputs in a high-dimensional or even infinite space. Support vectors are the values that are closest to the categorization margin. The purpose of the SVM is to increase the margin between the hyperplane and the support vectors as much as possible [1, 2].The core idea of SVM is to find a maximum marginal hyperplane that best divides the dataset into classes.

```
from sklearn import datasets
cancer = datasets.load_breast_cancer()
print("Features: ", cancer.feature_names)
# print the label type of cancer('malignant' 'benign')
print("Labels: ", cancer.target_names)
# print data(feature)shape
#cancer.data.shape
#print(cancer.data[0:5])
# print the cancer labels (0:malignant, 1:benign)
#print(cancer.target)
# Import train_test_split function
from sklearn.model_selection import train_test_split

# Split dataset into training set and test set
X_train, X_test, y_train, y_test = train_test_split(cancer.data, c
ancer.target, test_size=0.30,random_state=109) # 70% training and
30% test

#Import svm model
from sklearn import svm

#Create a svm Classifier
clf = svm.SVC(kernel='linear') # Linear Kernel

#Train the model using the training sets
clf.fit(X_train, y_train)

#Predict the response for test dataset
y_pred = clf.predict(X_test)

#Import scikit-learn metrics module for accuracy calculation
from sklearn import metrics

# Model Accuracy: how often is the classifier correct?
print("Accuracy:",metrics.accuracy_score(y_test, y_pred))

# Model Precision: what percentage of positive tuples are labeled
as such?
print("Precision:",metrics.precision_score(y_test, y_pred))

# Model Recall: what percentage of positive tuples are labelled as
 such?
print("Recall:",metrics.recall_score(y_test, y_pred))
```

```
Features:  ['mean radius' 'mean texture' 'mean perimeter' 'mean
area'
 'mean smoothness' 'mean compactness' 'mean concavity'
 'mean concave points' 'mean symmetry' 'mean fractal dimension'
 'radius error' 'texture error' 'perimeter error' 'area error'
 'smoothness error' 'compactness error' 'concavity error'
 'concave points error' 'symmetry error' 'fractal dimension error'
 'worst radius' 'worst texture' 'worst perimeter' 'worst area'
 'worst smoothness' 'worst compactness' 'worst concavity'
 'worst concave points' 'worst symmetry' 'worst fractal
dimension']
Labels:  ['malignant' 'benign']
Accuracy: 0.9649122807017544
Precision: 0.9811320754716981
Recall: 0.9629629629629629
```

5.4 Linearity with Kernel Methods

Kernels or kernel techniques (also known as kernel functions) are a collection of distinct sorts of pattern analysis algorithms. They are used in conjunction with a linear classifier to tackle a non-linear issue. Support vector machines (SVMs), which are utilized in classification and regression issues, utilizes kernel methods. The SVM employs a "kernel trick," in which the data are processed and an optimum boundary for the various outputs is discovered.

References

1. Durkin, A., Millan-Agorio, M., Guo, M., Process Systems Design Framework for Resource Recovery from Wastewater, in: *Computer Aided Chemical Engineering*, vol. 48, S. Pierucci, F. Manenti, G.L. Bozzano, D. Manca (Eds.), pp. 1039–1044, Elsevier, Department of Chemical Engineering, Imperial College London, SW7 2AZ, UK, 2020, ISSN 1570-7946, ISBN 9780128233771, https://doi.org/10.1016/B978-0-12-823377-1.50174-9,(https://www.sciencedirect.com/science/article/pii/B9780128233771501749).
2. Gove, R. and Faytong, J., Chapter 4 - Machine Learning and Event-Based Software Testing: Classifiers for Identifying Infeasible GUI Event Sequences, in: *Advances in Computers*, vol. 86, A. Hurson and A. Memon (Eds.), pp. 109–135, Elsevier, Human–Computer Interaction Lab, Department of Computer Science, University of Maryland, College Park, MD, USA, 2012, ISSN 0065-2458, ISBN 9780123965356, https://doi.org/10.1016/B978-0-12-396535-6.00004-1, (https://www.sciencedirect.com/science/article/pii/B9780123965356000041).
3. https://www.analyticsvidhya.com/blog/2020/01/fundamentals-deep-learning-activation-functions-when-to-use-them/.
4. https://www.codecademy.com/learn/perceptrons-and-neural-nets-skill-path/modules/perceptrons-skill-path/cheatsheet.

6

Distance-Based Models

6.1 Introduction

A machine learning approach called nearest neighbor classification seeks to identify previously unknown query items while differentiating two or more destination classes [1]. It discovers the nearest neighbor to an unknown pattern and gives the class name of this nearest neighbor to the unknown pattern [2]. A simple expansion of number of navigational contexts (NNC) is to consider the k-closest neighbors, with the class label determined by a majority vote [3].

As an example, consider a dataset consisting three features and a class label as shown in Table 6.1 with 10 objects in the dataset. The unknown object, Q(1.75, 1.75, 1.75), class label is identified by computing Euclidean distance of the unknown object to 10 objects. The computed Euclidean distance is shown in Table 6.2. The unknown object class label is class label of the object with least distance, i.e., the class label of Q is 1. The rule used here is known as NNC or 1-NNC because only one nearest neighbor is taken for classification. One can consider more than one neighbor called as k-NN classification model, where k is a positive integer [4, 5].

K-nearest neighbor is a strong and simple machine learning classification algorithm. This supervised learning approach has a wide range of applications, including pattern recognition, data mining, and prediction analysis. It is widely utilized in real-world situations because it is non-parametric, which means it does not make any assumptions about data distribution (as opposed to other algorithms such as Gaussian mixture model (GMM), which assume a Gaussian distribution of the given data). Prior data are supplied (also known as training data), which splits coordinates into groups based on an attribute [10].

Apart from Euclidean distance, other distance measures used for computation of distance for greater accuracies are hamming distance, Manhattan distance, and Minkowski distance [11].

Kolla Bhanu Prakash. Machine Learning for Industrial Applications, (87–102) © 2024 Scrivener Publishing LLC

Table 6.1 Sample dataset.

No	A	B	C	Class
x1	1	1	1	1
x2	2	1	3	1
x3	2	3	1	2
x4	2	2	1	1
x5	2	2	2	1
x6	2	3	1	2
x7	3	1	1	1
x8	4	2	2	1
x9	5	2	2	2
x10	4	2	1	2

Table 6.2 Computed Euclidean distance.

No	Euclidean distance
x1	1.29904
x2	1.47902
x3	1.47902
x4	0.82916
x5	0.43301
x6	1.47902
x7	1.63936
x8	2.27761
x9	3.26917
x10	2.38485

The issues of NNC are as follows:

a) It must save every single pattern in the training set, i.e., the full training set must be saved. If the training set size is small, then this appears to be a good idea, but, if the training set size is high, then this appears to be a bad idea [6].
b) NNC is afflicted by the curse of dimensionality. The feature set of a classifier is critical because it distinguishes between patterns. A good classifier has a number of different qualities. As the number of features increases, the classifier's performance deteriorates. The dimensionality issue is discussed in depth in [7, 8].
c) The third thing to examine is the outlier effect. An outlier is an unfavorable item in the training set. Outliers have an influence on classifiers that use distance measures as their primary metric. The influence of outliers on the sample set size is discussed in [9].

6.1.1 Distance-Based Clustering

Clustering techniques are essential components of unsupervised learning and machine learning in general. These algorithms assist locate structure in chaos by giving meaning to data that are not labeled. However, not all clustering algorithms are made equal, and each has its own set of advantages and disadvantages [12].

Image processing, neurology, economics, network communication, medicine, recommendation systems, and consumer segmentation are just a few of the applications that use cluster analysis. In addition, when working with a fresh dataset, clustering may be regarded the first step in extracting insights and understanding the data distribution. Cluster analysis may also be used to reduce the number of dimensions in a dataset (e.g., principal component analysis (PCA)). It might also be used as a preprocessing or intermediate step in other algorithms such as classification, prediction, and data mining [13].

6.2 K-Means Algorithm

Each partition represents a cluster, and n data items are divided into k partitions (k << n). At least one data point must be included in each cluster. Furthermore, each data object can only belong to one group. Observations from the same cluster should be comparable or within an acceptable range. Separate groups of things, on the other hand, must be distinguished. To put it another way, the k-means approach aims to maximize the distance between cluster centers while minimizing the distance between observations within each cluster [13].

K-means clustering works properly if the distribution's variance of every characteristic is spherical. Clusters are linearly separable, clusters have comparable numbers of observations, and variables present the equal variance [13].

It all starts with a random placement of k points in the features space, each of which represents a unique cluster's centroid. Iteratively calculate the distance between each dataset observation and each cluster center using a dissimilarity metric. Assign each observation to the cluster that is closest to the centroid. Calculate the mean (numeric attributes) for each cluster's points and reassign the centroid to the resultant mean for each cluster. This cycle will continue until a predetermined convergence criterion is met [13].

Given that their shapes are non-spherical, K-means believes that point "B" is closer to point "A" than point "C." Points "A" and "B" will therefore be in the same cluster; however, point "C" will be in a distinct cluster. Because it does not segregate related points, the single linkage hierarchical clustering approach gets this right (note).

K-means clustering is one of the most widely used clustering methods, and practitioners frequently use it as their first line of defense when tackling clustering problems to determine the general structure of the dataset. Data points are categorized using k-means into unique, non-overlapping groupings. When the clusters are somewhat spherical in shape, it works quite well. It suffers, nevertheless, because cluster geometric shapes diverge from spherical shapes.

Additionally, it requires a predefined number of clusters rather than learning it from the data. Knowing the underlying assumptions of algorithms and approaches will help you become a better practitioner by giving you a better understanding of the advantages and disadvantages of each technique. This will assist you in deciding when and under what conditions to apply each technique.

The advantages and disadvantages are as follows [13]:

Advantages:

i) the learning curve is relatively steep;
ii) widely implemented;

iii) fast convergence for clustering small datasets;
iv) easy to implement.

Disadvantages:

i) computationally expensive for large datasets(k becomes large);
ii) sometimes, it is difficult to choose an initial value for the number of clusters (k);
iii) does not guarantee to converge to a global minimum; it is sensitive to the centroids' initialization; different setups may lead to different results;
iv) strong sensitivity to outliers;
v) works only on numerical data;
vi) fails to give good quality of clustering for a group of points that have non-convex shapes.

6.2.1 K-Means Algorithm Working Process

The following stages illustrate how the K-means algorithm functions:

1. To determine the number of clusters, choose K.
2. Pick K centroids or random spots. (It might not be the supplied dataset.)
3. Assign each data point to its nearest centroid, which will create the K clusters that have been predetermined.
4. Place a new centroid for each cluster after computing the variance.

Figure 6.1 represents the variations before and after applying k-means algorithm.

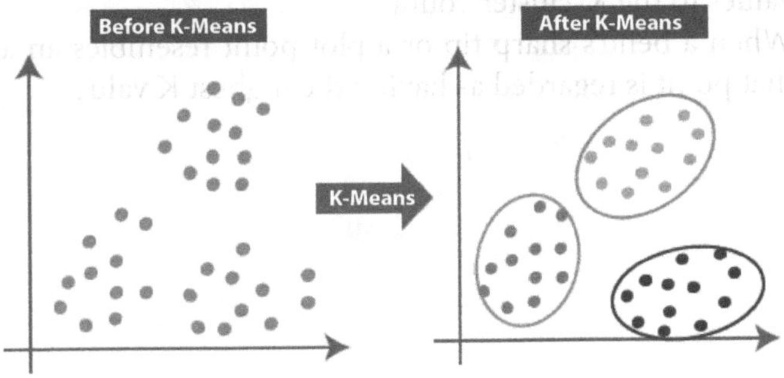

Figure 6.1 Before vs. after K-means.

5. Re-assign each data point to the new centroid of each cluster by repeating the third step.
6. If there is a reassignment, then proceed to step 4; otherwise, proceed to finish.
7. The model is ready.

The K-means clustering algorithm's effectiveness rests on the extremely effective clusters that it creates. However, determining the ideal number of clusters is a difficult process. There are several other approaches to determining the optimal number of clusters, but, in this article, we focus on the best technique. We use elbow method to find K.

6.3 Elbow Method

One of the most widely used techniques for determining the ideal number of clusters is the elbow method. The WCSS value idea is used in this technique. The term "total variations inside a cluster" is abbreviated as "WCSS," which stands for Within Cluster Sum of Squares [16].

We can use any method, such as the Manhattan distance or the Euclidean distance, to calculate the separation between the data points and the centroid.

The elbow approach goes through the procedures listed below to determine the best value of clusters:

1. On a given dataset, it does K-means clustering for various K values (ranges from 1 to 10).
2. Calculate the WCSS value for each value of K.
3. Trace a curve from the estimated WCSS values to the K-cluster count. Trace a curve from the estimated WCSS values to the K-cluster count.
4. When a bend's sharp tip or a plot point resembles an arm, that point is regarded as having the highest K value.

```
import pandas as pd
import numpy as np
import matplotlib.pyplot as plt
from sklearn.cluster import KMeans

def cluster_details(cnt):
  tot = 0
  for i, j in cnt.items():
    print('Cluster {0} has {1} items'.format(i,j))
    tot = tot + j
  print('Total Objects considered for clustering : %d'%tot)

def get_required_clusters():
  n = int(input('No of Clusters Required:'))
  return n

def main():
  a1=10*np.random.rand(100,2)

  n_clusters=get_required_clusters()
  kmean=KMeans(n_clusters)

  kmean.fit(a1)

  print('Cluster Centers:')
  print(kmean.cluster_centers_)

  print("Labeled Cluster Items:")
  klbl = list( kmean.labels_)
  print(klbl)
  l1 = [ i for i in range(n_clusters)]

  print('Cluster Labels:')
  print(l1)

  cnt = {}
  for i in l1:
    x = klbl.count(i)
    if i not in cnt:
      cnt[i] = x

  print('Cluster Details:')
  cluster_details(cnt)

main()
```

Python Code for K-Means

Output

```
No of Clusters Required:5
Cluster Centers:
[[6.89791756 8.22724599]
 [2.26299205 3.32537435]
 [8.70731584 3.06101648]
 [1.89892425 7.40880996]
 [5.74663979 1.91236991]]
Labeled Cluster Items:
[2, 0, 2, 0, 3, 0, 1, 3, 0, 3, 0, 0, 4, 1, 0, 2, 3, 2, 2, 3, 4, 1, 1,
3, 0, 1, 2, 0, 2, 0, 2, 3, 0, 0, 3, 0, 4, 1, 1, 0, 0, 2, 0, 4, 2, 0,
1, 4, 3, 3, 1, 3, 2, 2, 3, 4, 4, 3, 2, 1, 1, 2, 0, 3, 3, 0, 2, 0, 4,
0, 2, 0, 0, 0, 3, 3, 2, 3, 0, 3, 1, 4, 4, 4, 1, 4, 1, 1, 2, 0, 2, 4,
3, 0, 1, 2, 3, 2, 0, 2]
Cluster Labels:
[0, 1, 2, 3, 4]
Cluster Details:
Cluster 0 has 28 items
Cluster 1 has 16 items
Cluster 2 has 22 items
Cluster 3 has 21 items
Cluster 4 has 13 items
Total Objects considered for clustering : 100
```

6.4 K-Median

The median, which indicates the center point where other observations are uniformly scattered around it, is used in a modified version of the k-means method. In comparison to the mean, the median is less susceptible to outliers. The Manhattan distance is also used as a measure for calculating distances between observations. In addition, the method seeks to minimise the cost function.

K-median uses a new dissimilarity metric in an effort to reduce the susceptibility of k-means to outliers. We often utilize the absolute difference, also known as the L1 norm, the Manhattan, or taxicab distance, in place of the Euclidean distance.

6.4.1 Algorithm

Input parameter K (clusters) and iter (iterations)
Initialize random K objects in the dataset as centers
Do iter times:

Assign each item its nearest center.
Find new center

6.5 K-Medoids, PAM (Partitioning Around Medoids)

A medoid indicates a data point with the lowest average dissimilarity across all points inside a cluster in a modified version of the k-means method. The goal is to keep each cluster's overall cost as low as possible. It employs a medoid as a measure to reassign the centroid of each cluster, unlike k-means. Outliers have less of an impact on medoids. In contrast to k-means, these medoids represent real observations from the dataset rather than calculated points (mean value). The Manhattan distance is recommended as a measure because it is less susceptible to outliers [13].

Recall that we used centroids while computing K-means. The link between means and medians and that between centroids and medoids are similar. Centroids and means are almost always one of the data observations, although medians and medoids are not usually.

The key distinction between K-means and K-medoids is that, whereas K-medoids build clusters based on the distance to medoids, K-means form clusters based on the distance of observations to each centroid.

6.5.1 Advantages

- In the presence of outliers, more resilient than k-means.
- It is simple to implement.
- It converges in a predetermined amount of iterations and works well with a small dataset.

6.5.2 Drawbacks

- For a big dataset, it does not scale well.
- The cost of computational complexity is high.
- The value of the parameter k must be set.
- Convergence to a global minimum is not guaranteed. It is affected by the initialization of the centroids.
- The outcomes vary depending on the setting.
- It is solely capable of working with numbers.

6.5.3 Algorithm

There are many other kinds of algorithms for k-medoids clustering, but Partitioning Around Medoids, or PAM for short, is the easiest and most effective one. The following procedures are used in PAM to locate cluster centers [17]:

1. Pick k data points from the scatter plot to serve as the cluster's origin.
2. Calculate their distance from all the points in the scatter plot.
3. Put each point in the cluster whose center it is most closely associated with.
4. Choose a new point in each cluster that has a minimum distance from itself of all the points in that cluster.
5. Repeat *Step 2* until the centers stop changing.

```
Import pandas as pd
import numpy as np
import matplotlib.pyplot as plt

from sklearn_extra.cluster import KMedoids

def cluster_details(cnt):
  tot = 0
  for i, j in cnt.items():
    print('Cluster {0} has {1} items'.format(i,j))
    tot = tot + j
  print('Total Objects considered for clustering : %d'%tot)

def get_required_clusters():
  n = int(input('No of Clusters Required:'))
  return n

def main():
  a1=10*np.random.rand(100,2)

  n_clusters=get_required_clusters()
  kmedoids=KMedoids(n_clusters)

  kmedoids.fit(a1)

  print('Cluster Centers:')
  print(kmedoids.cluster_centers_)

  print("Labeled Cluster Items:")
  klbl = list( kmedoids.labels_)
  print(klbl)
  l1 = [ i for i in range(n_clusters)]

  print('Cluster Labels:')
```

```
    print(l1)

    cnt = {}
    for i in l1:
      x = klbl.count(i)
      if i not in cnt:
        cnt[i] = x

    print('Cluster Details:')
    cluster_details(cnt)

main()
```

Output

```
No of Clusters Required:5
Cluster Centers:
[[1.63801097 4.40737169]
 [5.53597263 1.05185474]
 [4.91567767 8.53633198]
 [7.75493982 4.24147202]
 [9.13887038 7.58357196]]
Labeled Cluster Items:
[1, 2, 2, 2, 3, 3, 3, 2, 1, 2, 1, 3, 2, 4, 4, 4, 0, 3, 1, 3, 1, 0, 0,
0, 2, 4, 0, 0, 0, 1, 3, 1, 4, 2, 3, 0, 0, 2, 4, 4, 2, 1, 3, 3, 2, 4,
0, 1, 4, 4, 1, 1, 2, 3, 1, 0, 2, 4, 0, 4, 1, 0, 2, 2, 1, 1, 1, 1, 0,
1, 4, 0, 0, 1, 4, 0, 0, 0, 0, 0, 3, 2, 3, 3, 2, 4, 0, 2, 1, 0, 2, 2,
3, 3, 2, 2, 0, 1, 4, 0]
Cluster Labels:
[0, 1, 2, 3, 4]
Cluster Details:
Cluster 0 has 25 items
Cluster 1 has 21 items
Cluster 2 has 22 items
Cluster 3 has 16 items
Cluster 4 has 16 items
Total Objects considered for clustering : 100
```

For efficiency improvement of PAM, the CLARA algorithm is used.

6.6 CLARA (Clustering Large Applications)

It is a sample-based strategy that picks a small group of data points at random rather than examining all of the observations; therefore, it works well with huge datasets. Furthermore, k-medoids are selected from the previously specified sample. PAM's scalability will be improved as a result of this (reduce computing time and memory allocation problem). It works on multiple batches of the dataset in order to obtain the best outcome [13].

The following method differs from PAM and CLARA in that it is based on sampling. Only a small portion of the actual data is selected as a representative of the data, and, using PAM, medoids are selected from this sample.

The theory holds that, if the sample is picked fairly randomly, then it will accurately represent the entire dataset, and, as a result, the representative items (medoids) selected will be comparable to those chosen from the entire dataset.

6.6.1 Advantages

- The ability to work with enormous datasets.
- When working with a huge data collection, reduce the computation time.
- The ability to deal with outliers.

6.6.2 Disadvantages

- The sample size and the value of k have an impact on efficiency.
- The quality of clustering is determined by the sampling strategy employed.
- It is difficult to put into practise.

6.7 CLARANS (Clustering Large Applications Based on Randomized Search)

It is advancement of k-medoid. It is more effective than PAM and CLARA.

By merely examining a portion of the dataset and not being restricted to any particular sample at any one time, the CLARANS method combines the PAM and CLARA algorithms. While CLARANS draws a sample with some randomness at each stage of the search, CLARA always uses a constant sample.

The disadvantages of the K-medoids and CLARA algorithms are addressed by the CLARANS algorithm, which also handles geographical data, a challenging type of data mining data. It keeps the cost of computation and the impact of data sampling on cluster formation in check.

6.7.1 Advantages

- The ability to deal with outliers.

6.7.2 Disadvantages

- Implementation is difficult.
- The quality of clustering is determined by the sampling strategy employed.

6.7.3 Algorithm

1. Choose "k" random data points and designate them as medoids for now.
2. Choose a point at random, say "a," from among the points chosen in step 1 and another point, say "b," which is not among them.
3. The calculation needed to choose the points in step would have already been done; thus, we would have the total distance between point "a" and all other points (1). Similar calculations should be made for point "b."
4. Replace "a" with "b" if the sum of the distances from all other places for point "b" turns out to be smaller than that for point "a."
5. The number of local minima computed, or the number of iterations to be carried out, is specified as a parameter, and the method executes such a randomized search of medoids "x" times. The phrase "local optimum" refers to the set of medoids that were reached after taking such "x" steps [18].
6. Every time a point replacement is accomplished, a counter is increased. Until the counter does not exceed the maximum number of neighbors to be reviewed, the points are examined again for replacement.
7. The best local optimum selection of medoids is the set of medoids that the algorithm produces when it finishes.

6.8 Hierarchical Clustering

Hierarchical clustering is the process of dividing data into groups based on some measure of similarity, evaluating how they are alike and different, and narrowing down the data. Hierarchical clustering is divided as agglomerative and divisive clustering.

Therefore, as we have seen in the K-means clustering, this algorithm has some drawbacks, including a preset number of clusters and a tendency to always try to generate clusters of the same size. Because we do not need to be aware of the specified number of clusters while using the hierarchical clustering technique, we can choose this algorithm to address these two problems.

There are two approaches in hierarchical clustering. These are as follows:

1. Agglomerative: Agglomerative algorithms start by treating all data points as single clusters and then combine them until only one cluster is left. This method is bottom-up.
2. Divisive: Because it takes a top-down approach, the divisive algorithm is the opposite of the agglomerative algorithm.

6.9 Agglomerative Nesting Hierarchical Clustering (AGNES)

This is a bottom-up strategy in which the algorithm begins by grouping all data points into single clusters and merging them until only one remains.

The strategies are necessary for determining whether objects/clusters should be combined or divided based on item similarity [14].

There are several mechanisms for cluster agglomeration (i.e., linkage methods). The most prevalent ways of connection are given here [14].

- Maximum or complete linkage: The greatest value of all pairwise distances between items in cluster 1 and elements in cluster 2 is used to calculate the distance between two clusters. It results in clusters that are more compact.
- Minimum or single linkage: The minimum value of all pairwise distances between items in cluster 1 and elements in cluster 2 is used to calculate the distance between two clusters. Long, "loose" clusters are a common result.
- Mean or average linkage: The average distance between the items in cluster 1 and cluster 2 is used to calculate the distance between two clusters.
- Centroid linkage: The distance between two clusters is measured between cluster 1's centroid (a mean vector of length p variables) and cluster 2's centroid.
- Ward's minimum variance method: It reduces the overall within-cluster variation to a minimum. At each phase, the clusters with the shortest distance between them are merged.

The basic process of hierarchical clustering is as below [15].

Step 1: Begin by allocating each item to a cluster, such that, if there are n things, then there will be n clusters, each with one item.
Step 2: Merge the closest (most similar) set of clusters into a single cluster, resulting in one fewer cluster.
Step 3: Calculate the distances (likelihoods) between the new and old clusters.
Step 4: Repeat steps 2 and 3 until all of the items are clustered into a single size n cluster.

6.10 DIANA

DIANA is also known as divisive analysis clustering algorithm. The dividing method is the inverse of the agglomerative algorithm in that it uses a top-down technique to divide a single cluster into n clusters, each having at least one item. The clusters are divided into two groups based on their degree of similarity. This is done in a recurrent fashion until separate cluster groups emerge.

We can claim that the agglomerative hierarchical clustering is exactly what the divisive hierarchical clustering is not. In divisive hierarchical clustering, all of the data points are considered as a single cluster, and, in each iteration, the data points that are not comparable are separated from the clusters. N clusters are all that are left in the end.

References

1. Seidl, T., Nearest Neighbor Classification, in: *Encyclopedia of Database Systems*, L. Liu and M.T. Özsu (Eds.), Springer, Boston, MA, 2009, https://doi.org/10.1007/978-0-387-39940-9_561.
2. Cover, T. and Hart, P., Nearest neighbor pattern classification. *IEEE Trans. Inf. Theory*, 13, 1, 21–27, 1967.
3. Babu, V. S. and Viswanath, P., Weighted k-nearest leader classifier for large data sets, in: *International Conference on Pattern Recognition and Machine Intelligence*, pp. 17–24, Springer, Berlin Heidelberg, 2007, December.
4. Raja Kumar, R., Viswanath, P., ShobhaBindu, C., Nearest Neighbor Classifiers: A Review. *Int. J. Comput. Intell. Res.*, 13, 2, 303–311, 2017, ISSN 0973-1873 Available at https://www.ripublication.com/ijcir17/ijcirv13n2_13.pdf.

5. Sun, B. and Chen, H., A Survey of Nearest Neighbor Algorithms for Solving the Class Imbalanced Problem. *Wirel. Commun. Mobile Comput.*, 2021, 12, 2021, Article ID 5520990, https://doi.org/10.1155/2021/5520990.
6. Devi, V. S. and Murty, M. N., An incremental prototype set building technique. *Pattern Recognit.*, 35, 2, 505–513, 2002.
7. Spruyt, V., The Curse of Dimensionality in classification, http://www.visiondummy.com/2014/04/curse-dimensionality-affectclassification/ [Last accessed 29.12.2021].
8. Keogh, E. and Mueen, A., Curse of dimensionality, in: *Encyclopedia of Machine Learning*, pp. 257–258, Springer US, University of California-Riverside, Riverside, CA, USA, 2011.
9. Selst, M.V. and Jolicoeur, P., A solution to the effect of sample size on outlier elimination. *Q. J. Exp. Psychol.*, 47, 3, 631–650, 1994.
10. Srivastava, T., Introduction to k-Nearest Neighbors: A powerful Machine Learning Algorithm (with implementation in Python & R), https://www.analyticsvidhya.com/blog/2018/03/introduction-k-neighbours-algorithm-clustering/ Last Accessed on 29.12.2021.
11. Brownlee, J., 4 Distance Measures for Machine Learning, https://machinelearningmastery.com/distance-measures-for-machine-learning/. {Last Accessed on 29.12.2021].
12. Iliassich, L., Clustering Algorithms: From Start To State Of The Art, https://www.toptal.com/machine-learning/clustering-algorithms [Last Accessed on 29.12.2021].
13. Harmouch, M., 17 Clustering Algorithms Used In Data Science and Mining - An overview of clustering algorithms, their use cases, and their advantages and disadvantages, https://towardsdatascience.com/17-clustering-algorithms-used-in-data-science-mining-49dbfa5bf69a, [Last Accessed on 29. 12.2021].
14. Kassambara, A., HIERARCHICAL CLUSTERING IN R: THE ESSENTIALS, https://www.datanovia.com/en/lessons/agglomerative-hierarchical-clustering/, [Last Accessed on 30.12.2021].
15. Gan, G., Ma, C., Wu, J., *Data Clustering: Theory, Algorithms, and Applications*, ASA-SIAM, VA, 2007.
16. https://www.javatpoint.com/k-means-clustering-algorithm-in-machine-learning.
17. https://subscription.packtpub.com/book/big-data-and-business-intelligence/9781789956399/1/ch01lvl1sec08/introduction-to-k-medoids-clustering.
18. https://analyticsindiamag.com/comprehensive-guide-to-clarans-clustering-algorithm/.

7
Model Ensembles

Machine learning algorithms have limitations, making it difficult to create a high-accuracy model [1]. A single algorithm may not be able to deliver the best forecast for a given dataset [2]. Ensemble modeling is a method of predicting a result by combining several different base models. The goal of employing ensemble models is to lower the prediction's generalization error. When using the ensemble approach, the prediction error lowers as long as the basis models are diverse and independent [3]. It is a solution to overcome the technical challenges like high variance, low accuracy, and features of noise and bias in building a single estimator [2]. Ensemble approaches in machine learning have risen in prominence as a result [4]. The same is graphically presented in Figure 7.1.

The main goal of ensemble learning is to enhance the performance of models in areas like classification, prediction, and function approximation. We can summarize ensemble learning in the following straightforward manner:

A machine learning model called an ensembled model integrates the forecasts from two or more models.

The most popular ensemble methods are as follows [4]:

- Bagging
- Boosting
- Stacking

Each strategy is described by an algorithm, but their success has generated a slew of extensions and related ideas. As a result, it is better to think of them as a set of methodologies or conventional approaches to ensemble learning [5].

Figure 2 represents ensemble types.

Kolla Bhanu Prakash. Machine Learning for Industrial Applications, (103–116) © 2024 Scrivener Publishing LLC

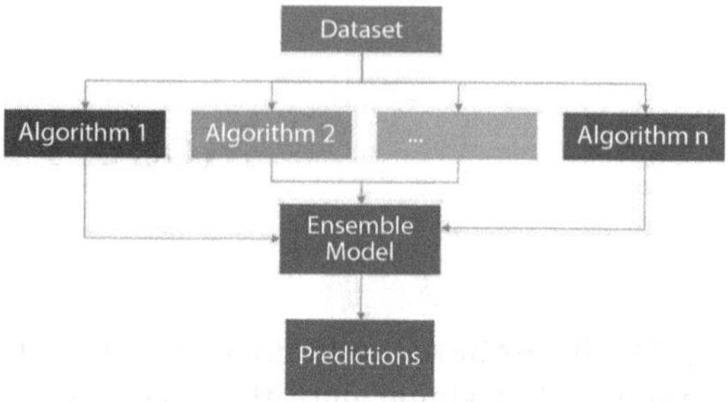

Figure 7.1 Ensemble model for prediction using multiple algorithms.

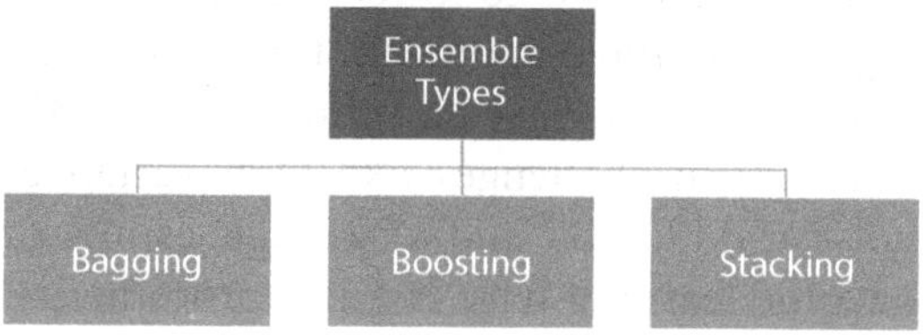

Figure 7.2 Ensemble types.

7.1 Bagging

Bootstrap aggregation, often known as bagging, is an ensemble learning method that uses varied training data to find a diverse collection of ensemble members [5]. It is a technique for merging the outputs of various models (for example, all decision trees) to produce a more generic result [6]. It is applied classification and regression [4]. The process is depicted in the Figure 7.3.

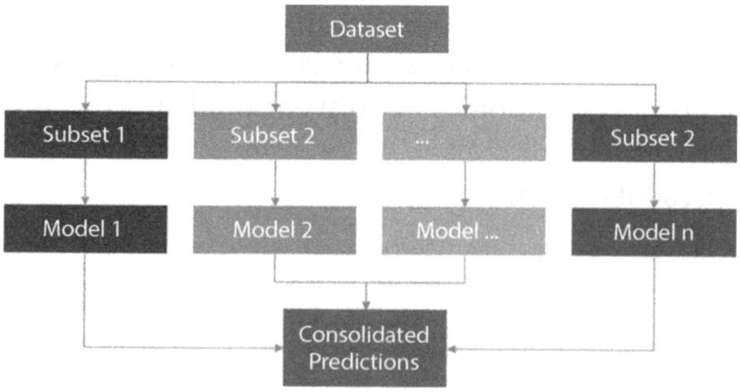

Figure 7.3 Bagging.

Bagging is the process of creating several subsets from a single dataset by replacing observations. The ability to duplicate sample instances in a set is enabled via replacement. Each subset is of the same size and may be used to train models in the same way [2]. For each of these subsets, a base model or a weak model is produced. The models are independent of one other and run in parallel. All of the models' forecasts are combined to produce the final predictions [6]. Bagging reduces variance and minimizes overfitting [2].

Random forest builds multiple split trees using a subset of training samples and characteristics. To fit each training set, multiple decision trees are created. The distribution of samples/features is usually done at random. The replication of sample instances in a set is possible with replacement. Each subset is of uniform size and can be used to train models [2].

The basic concept behind bagging is to fit a number of different models and "average" their forecasts to get a model with a lower variance. In reality, we are unable to construct entirely independent models because doing so would necessitate an excessive amount of data. Therefore, in order to fit almost independent models, we rely on the strong "approximate qualities" of bootstrap samples (representativity and independence).

In order for each new bootstrap sample to function as another (nearly) independent dataset derived from the true distribution, we first produce a number of bootstrap samples. Finally, we may aggregate them such that we sort of "average" their outputs and get an ensemble model with less variance than its components by fitting a weak learner to each of these samples. The learned base models are roughly independent and identically distributed in the same way that the bootstrap samples are. Then, "averaging" the outputs of inexperienced learners does not alter the expected result but does lower its variance.

The several models fitted in parallel can be combined in a variety of ways. The output of the ensemble model for a regression problem can be produced by arithmetically averaging the results of the individual models. The class outputted by each model for a classification problem can be thought of as a vote, and the ensemble model returns the class that obtains the majority of the votes (this is called hard voting). We can, nevertheless, take into account the probabilities of each class supplied by each model for a classification task, average these probabilities, and keep the class with the highest average probability (this is called soft-voting). If any appropriate weights may be applied, averages or votes can either be simple or weighted.

The random forest approach is a bagging technique where deep trees along with an output with lower variance are fitted on bootstrap samples. But random forests also employ another tactic to reduce the correlation between the numerous fitted trees: when growing each tree, we sample over features as well as observations in the dataset to create a bootstrap sample, keeping just a random subset of them to build the tree.

In fact, sampling over features has the effect of preventing all trees from using the exact same data to guide their decisions, which lowers the correlation between the many returning outputs. The decision-making process is more resilient to missing data when sampling is used instead of features, as observations (whether from the training dataset or not) with missing data can still be regressed or classed using trees that include only features when data are present. In order to produce more reliable models, the random forest algorithm combines the ideas of bagging and random feature subspace selection.

7.1.1 Advantages

- Easy Implementation: The performance of models can be increased by combining the predictions of base learners or estimators using Python tools like scikit-learn (also known as sklearn).
- Variance Reduction: A learning algorithm's variance can be decreased via bagging. This is especially useful for high-dimensional data, where the presence of missing values can increase variance, increase the risk of overfitting, and hinder proper generalization to new datasets.

7.1.2 Disadvantages

- Interpretability loss: Because of the averaging that occurs among forecasts, it is challenging to derive extremely precise business insights from bagging. A more accurate or full dataset could potentially produce more accuracy inside a single classification or regression model, even though the output is more exact than any particular data point.
- Expensive Computation: With more iterations, bagging slows and becomes more time-consuming. It is therefore inadequate for real-time applications. For fast-generating bagged ensembles on sizable test sets, clustered computers with a lot of processing cores are appropriate.

- Less Flexible: Bagging as a strategy is particularly effective with less reliable algorithms. Less variety exists within the dataset of the model; therefore, the ones that are more stable or highly biassed do not gain as much.

7.1.3 Bagging Workage [9]

Bootstrapping: Bagging uses a bootstrapping sampling method to provide a variety of samples. This resampling technique creates several subsets of the training dataset by randomly and replacement-wise picking data points. This indicates that you have the option to select the same instance more than once each time you choose a data point from the training dataset. A value or instance repeated twice (or more) in a sample as a result.

Parallel Training: These bootstrap samples are then trained separately and concurrently using base or weak learners.

Aggregation: Finally, an average or a majority of the predictions are picked, depending on the job (such as regression or classification), to produce a more accurate estimate. Regression involves averaging every output predicted by each classifier; this process is referred to as soft voting. Hard voting, often known as majority voting, is the process of accepting the class that receives the most number of votes in classification issues.

7.1.4 Algorithm

1. From the original dataset, many equal-sized subsets are made, choosing observations with replacement.
2. On each of these subsets, a foundation model is built.
3. Every model is independently and concurrently learned with every training set.
4. The combined forecasts from all the models yield the final results.

7.2 Boosting

Boosting is an ensemble strategy for improving future predictions by learning from previous predictor errors. The method combines numerous weak base learners to create a single strong learner, greatly boosting model predictability. Boosting works by placing weak learners in a sequence so that they can learn from the next learner in the sequence, resulting in more accurate predictions [4].

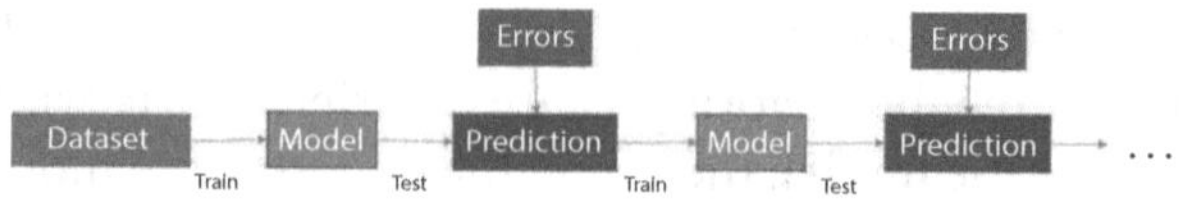

Figure 7.4 AdaBoost's sequential learning creates a more powerful learnt model.

Gradient boosting, adaptive boosting (AdaBoost), and XGBoost are all examples of boosting techniques (extreme gradient boosting). AdaBoost employs weak learners in the form of decision trees, which typically have one split, dubbed decision stumps. The primary decision stump in AdaBoost is made up of observations with equal weights [4]. It is shown in Figure 7.4.

Gradient boosting adds predictors to the ensemble in a sequential order, with prior predictors correcting their successors, resulting in a higher accuracy model. To compensate for the consequences of earlier predictor failures, new predictors are created. The gradient booster uses the gradient of descent to discover and correct errors in learners' predictions [4].

XGBoost improves speed and performance by employing decision trees with enhanced gradients. It is extremely dependent on the target model's computing speed and performance. Gradient boosted machines must be implemented in a sequential manner, which makes them slow to implement [4].

Similar to bagging approaches, boosting methods develop a family of models that are then combined to produce a strong learner that performs better. Contrary to bagging, which primarily aims to reduce variance, boosting is a technique that entails sequentially fitting a number of weak learners in a highly adaptive manner. Each model in the sequence is fitted while giving more weight to dataset observations that the previous models in the sequence handled poorly. It makes sense that each new model concentrates its efforts on the observations that have proven to be the most challenging to fit so far, in order to produce a strong learner with less bias. Boosting is a technique that, like bagging, may be applied to classification and regression problems.

The base models that are frequently taken into consideration for boosting are those with low variance but high bias because they are primarily focused on lowering bias. For instance, we would often select shallow decision trees with a limited number of depths if we were to use trees as our basic models. Another crucial factor that encourages the adoption of low variance but high bias models as weak learners for boosting is the fact that fitting these models typically requires less computing power (few degrees of freedom when parameterized). Indeed, fitting multiple complex models sequentially may become prohibitively expensive because calculations to fit the various models cannot be performed in parallel (unlike bagging).

We still need to specify how the weak learners will be sequentially fitted (what data from prior models do we take into account when fitting the current model?) and aggregated (how do we aggregate the current model to the prior ones?). In the two subsections that follow, we will go through these issues while focusing particularly on two crucial boosting techniques: AdaBoost and gradient boosting.

In essence, the way that these two meta-algorithms generate and aggregate the weak learners during the sequential process varies. While gradient boosting updates the value of these observations, adaptive boosting modifies the weights associated with each training dataset observation. The two approaches each use to determining the best model that can be represented as a weighted sum of weak learners are the primary source of this disparity.

7.2.1 Types of Boosting [7]

Boosting techniques concentrate on repeatedly merging weak learners to create a strong learner that can make more precise predictions. Recall that a weak learner classifies data just marginally better than chance. For tasks like image retrieval, this method can even outperform neural networks and support vector machines in terms of robust results.

Adaptive Boosting or Ada Boost: The AdaBoost algorithm was developed by Yoav Freund and Robert Schapire. In order to reduce the training error, this approach iteratively locates misclassified data points and modifies their weights. The model keeps improving in a sequential manner until the strongest predictor is produced.

Gradient Boosting: Gradient boosting is a technique created by Jerome H. Friedman and Leo Breiman that systematically adds predictors to an ensemble, with each one correcting for the errors of the previous one. As opposed to AdaBoost, which modifies the weights of individual data points, gradient boosting trains on the remaining errors from the prior predictor. Because it combines the gradient descent algorithm and the boosting method, the term "gradient boosting" is employed.

Extreme Gradient Boosting or XG Boost: A gradient boosting system created for speed and scale is called XGBoost. Through the use of the CPU's numerous cores, XGBoost enables parallel learning during training.

7.2.2 Advantages

- Easy Implementation: Several hyperparameter tweaking options can be combined with boosting to enhance fitting. Because boosting methods like contain built-in techniques to manage missing data, no data preprocessing is necessary. AdaBoost, XGBoost, and other well-known boosting techniques are simple to implement in Python thanks to the scikit-learn library of ensemble methods (also known as sklearn.ensemble).
- Bias Reduction: Boosting algorithms sequentially combine several weak learners, iteratively improving on observations. This method can aid in lowering excessive bias, which is frequently present in logistic regression models and shallow decision trees.
- Effective Computation: Boosting algorithms can aid in reducing dimensionality and improving computational efficiency because they only choose characteristics during training that improve their predictive value.

7.2.3 Disadvantages

- Overfitting: Whether boosting can lessen overfitting or make it worse is a matter of some debate. We classify it as a difficulty because, when it does happen, predictions cannot be extrapolated to fresh datasets.
- Intensive computation: Scaling sequential training in boosting is challenging. Boosting models can be computationally expensive because each estimator builds on its predecessors, but XGBoost aims to overcome scalability difficulties observed in other types of boosting approaches. When opposed to bagging, training boosting algorithms can take longer because a lot more parameters can affect the model's behavior.

7.2.4 Algorithm

1. Initialize the dataset and give each data point the same amount of weight.
2. Give the model this as input and find the data points that were incorrectly categorized.

3. Incorrectly classified data points should have more weight, whereas correctly classified data points should have less weight. Then, adjust all data points' weights to their original values.
4. if (got required results)
   Goto step 5
   else
   Goto step 2
5. End

7.3 Stacking

Stacked generalization, or stacking, is an ensemble approach for finding a varied group of members by shifting the model types fit on the training data and combining predictions with a model. Stacking has its own terminology, with level-0 models referring to ensemble members and level-1 models referring to the model that is used to integrate the forecasts [5]. The architecture is shown in Figure 7.5.

Although additional levels of models can be utilized, the most frequent strategy is a two-level hierarchy of models. Instead of a single level-1 model, we may have three or five level-1 models and a single level-2 model that integrates level-1 model predictions to generate a forecast [5].

Any machine learning model may be used to aggregate the predictions, although linear models, such as linear regression and logistic regression for binary classification, are the most common. The model's complexity may then be confined in lower-level ensemble member and simple models,

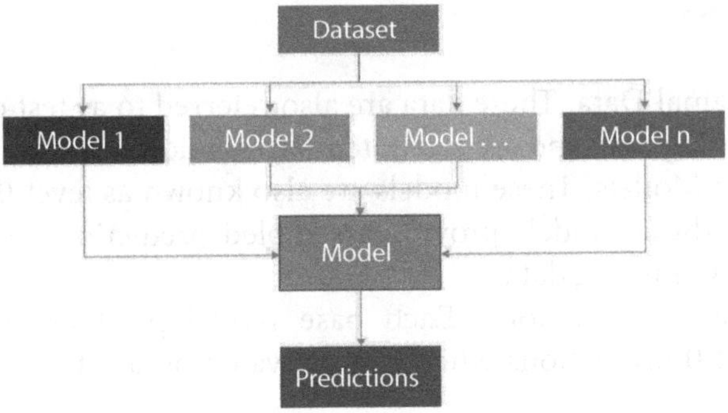

Figure 7.5 The final forecast is made using a stacking method in an ensemble architecture.

which can then learn to harness the range of predictions made. As ensemble members, the many machine learning models give variety [5].

As a consequence, it is ideal to use a collection of models that are taught or developed using a range of methodologies, ensuring that they make a variety of assumptions and, as a result, produce fewer prediction errors [5].

Stacking is an ensemble technique that teaches the model to integrate predictions made by learner models with predictions made by meta-models to create a final model with accurate predictions. The fundamental advantage of stacking ensembles is that they can protect a variety of effective models' abilities to address classification and regression issues. Additionally, it helps to develop a superior model with predictions that outperform all individual models. The many ensemble approaches and their definitions, the stacking ensemble method, the structure of stacking models, and the processes to implement stacking models in machine learning have all been covered in this topic.

As a more advanced version of the model averaging ensemble methodology, stacking is also referred to as a stacked generalization. In this method, all sub-models equally contribute based on their performance weights to create a new model with more accurate predictions. This new model is called stacking because it is piled on top of the older models.

7.3.1 Architecture of Stacking [8]

A meta-model that integrates the predictions of the base models makes up the stacking model's architecture, which consists of two or more base/ models or learners. The meta-model is referred to as the level-1 model, while these base models are referred to as level-0 models. In this way, the stacking ensemble approach incorporates the original (training) data, primary level models, primary level prediction, secondary level model, and final prediction.

- Original Data: These data are also referred to as test data or training data and are separated into n-folds.
- Base Models: These models are also known as level-0 models. These models provide assembled predictions (level 0) using training data.
- Level-0 Predictions: Each base model produces various level-0 predictions when it is activated on a set of training data.
- Meta-Model: One meta-model makes up the stacking model's architecture, which enables the best possible fusion of the

base models' predictions. The level-1 model is another name for the meta-model.
- Level-1 Prediction: When data not used to train the base models are fed to the meta-model, predictions are made, and these predictions, along with the expected outputs, provide the input and output pairs of the training dataset used to fit the meta-model. The meta-model learns how to best combine the predictions of the base models and is trained on different predictions made by individual base models.

7.3.2 Stacking Ensemble Family [8]

Other ensemble techniques exist that can be viewed as the predecessors of the stacking method. To make them easier to understand and to highlight the distinctiveness of each technique, we have separated them into the several frameworks of basic stacking. Let us talk about a few often utilized ensemble stacking strategies.

- Voting Ensembles: It is one of the simplest stacking ensemble approaches; it uses various algorithms to separately prepare each member. The voting ensemble, in contrast to the stacking technique, uses simple statistics rather than learning how to optimally aggregate predictions from base models independently. Regression problems that need us to estimate the mean or median of the base model predictions are important to tackle. Additionally, it aids in a variety of classification issues based on the overall number of votes collected for prediction. Hard voting refers to the label that receives a greater percentage of votes, and soft voting refers to the label that receives a higher percentage of probability votes or less votes.
- Weighted Average Ensemble: The voting ensemble, which makes use of a variety of model types as contributing members, is thought of as the next stage up. This technique calculates the average weight of each ensemble member based on their performance using some training datasets. Weighing each member according to how well it performs on a hold-out dataset, such as a validation set or out-of-fold predictions during k-fold cross-validation, is an improvement over this crude method. Additionally, it can entail adjusting the

coefficient weightings for each model based on the outcome of an optimization technique on a holdout dataset.

- Blending Ensemble: A similar strategy to stacking with a particular arrangement is blending. The process of preparing out-of-sample predictions for the meta-model using k-fold cross-validation is referred to as a stacking approach. With this approach, the training dataset is divided into several training sets and validation sets before learner models are trained using the training sets. Additionally, predictions are created using the validation set, sample set, and new model, which is then utilized to make final predictions on the test set utilizing the prediction values as features.
- Super Learning Ensemble: This technique and blending, which uses a certain arrangement of a stacking ensemble, are quite similar. It creates a meta-model using learner model out-of-fold predictions. It is still regarded as a modified form of blending, with the only distinction being in the method used to create out-of-sample predictions for the meta-learner.

Steps for implementing stack model:

1. Because creating training datasets for meta-models is the most typical method, divide training datasets into n-folds using the RepeatedStratifiedKFold.
2. The first fold, which is n – 1, has now been fitted to the base model, and it will now generate predictions for the nth folds.
3. The x1 train list is updated with the prediction given in the previous phase.
4. Steps 2 and 3 should be repeated for the remaining n – 1 folds to get an array of size n called x1 train.
5. The model has now been trained on all n parts and can now predict the results of the sample data.
6. Include this forecast in the y1 test list.
7. By using Models 2 and 3 for training, respectively, we may determine x2 train, y2 test, x3 train, and y3 test to obtain level-2 predictions.
8. Now, train the meta-model using level-1 predictions. The model will use these predictions as features.
9. Finally, a prediction on test data in the stacking model may now be made using meta-learners.

References

1. Kotu, V. and Deshpande, B., Chapter 2 - Data Mining Process, in: *Predictive Analytics and Data Mining*, V. Kotu and B. Deshpande (Eds.), pp. 17–36, Morgan Kaufmann, Burlington, Massachusetts, USA, 2015, ISBN 9780128014608, https://doi.org/10.1016/ B978-0-12-801460-8.00002-1. (https://www.sciencedirect.com /science/article/ pii/B9780128014608000021).
2. Alhamid, M., Ensemble Models A guide to learning ensemble techniques in a simple walkthrough, https://towardsdatascience.com /ensemble-models-5a62d4f4cb0c.
3. Kotu, V. and Deshpande, B., Chapter 2 - Data Science Process, in: *Data Science*, Second Edition, V. Kotu and B. Deshpande (Eds.), pp. 19–37, Morgan Kaufmann, Burlington, Massachusetts, USA, 2019, ISBN 9780128147610, https://doi.org/10.1016/B978-0-12-814761-0.00002-2. (https://www.sciencedirect.com/science/article/pii/B9780128147610000022).
4. CFI, Ensemble Methods-Combining multiple models to improve the desired results, https://corporatefinanceinstitute.com/resources/knowledge/other/ensemble-methods/.
5. Brownlee, J., A Gentle Introduction to Ensemble Learning Algorithms, https://machinelearningmastery.com/tour-of-ensemble-learning-algorithms/.
6. Singh, A., A Comprehensive Guide to Ensemble Learning (with Python codes), https://www.analyticsvidhya.com/blog/2018/06/comprehensive-guide-for-ensemble-models/.
7. https://towardsdatascience.com/ensemble-methods-bagging-boosting-and-stacking-c9214a10a205
8. https://www.javatpoint.com/stacking-in-machine-learning
9. https://www.ibm.com/cloud/learn/bagging

8

Binary and Beyond Binary Classification

Binary and beyond binary classifications are mainly classified as follows:

a) Classification
b) Scoring and ranking
c) Class probability estimation
d) Handling more than two classes
e) Descriptive learning

Descriptive learning theories make statements about how learning occurs and devise models that can be used to explain and predict learning results [1]. When describing different descriptive theories of learning below, we will follow the common categorization that distinguishes between behaviorist, cognitive, and constructive learning theories.

8.1 Binary Classification

A typical machine learning challenge is classifying data into one of two categories. You might need to foretell if a consumer is likely to make a purchase, whether a credit card transaction was fraudulent, whether deep space signals indicate the existence of a new planet, or whether a medical test indicates the presence of a disease. These are all difficulties of binary categorization.

Binary classification tasks with two class labels are referred to as binary classification.

Examples comprise the following:

- Detection of spam email (spam or not).
- Churn forecast (churn or not).
- Prediction of conversion (buy or not).

Kolla Bhanu Prakash. Machine Learning for Industrial Applications, (117–126) © 2024 Scrivener Publishing LLC

Binary classification problems often require two classes—one representing the normal state and the other representing the aberrant state.

For instance, the normal condition is "not spam," whereas the abnormal state is "spam." Another illustration is when a task involving a medical test has a normal condition of "cancer not identified" and an abnormal state of "cancer detected."

Class label 0 is given to the class in the normal state, whereas class label 1 is given to the class in the abnormal condition.

A model that forecasts a Bernoulli probability distribution for each case is frequently used to represent a binary classification task.

The discrete probability distribution known as the Bernoulli distribution deals with the situation where an event has a binary result of either 0 or 1. In terms of classification, this indicates that the model forecasts the likelihood that an example would fall within class 1 or the abnormal state.

A typical machine learning challenge is classifying data into one of two categories. You might need to foretell if a consumer is likely to make a purchase, whether a credit card transaction was fraudulent, whether deep space signals indicate the existence of a new planet, or whether a medical test indicates the presence of a disease. These are all difficulties of binary categorization.

The classes in your raw data may be represented by strings such as "Yes" and "No" or "Dog" and "Cat." We will first assign a class label before using this data; one class will be 0, and the other will be 1. The data are transformed into a form that a neural network.

Figure 8.1 represents classifying data in two classes.

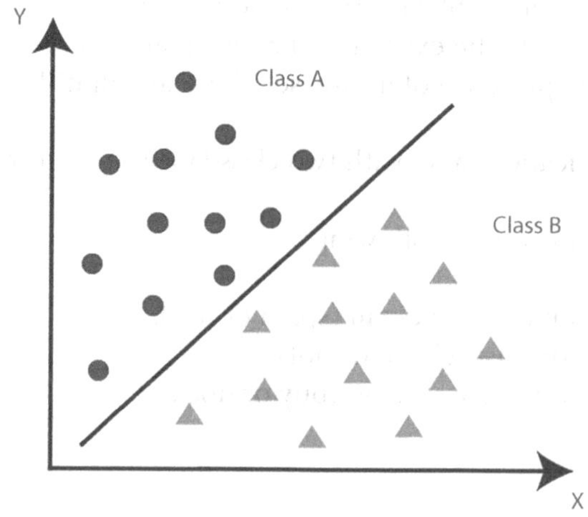

Figure 8.1 Classifying data in two classes (A and B).

A good example of binary classification is spam e-mail filtering, in which spam is typically seen as the positive class and ham as the negative class (obviously, positive here does not mean "good!").Medical diagnosis (where the positive class is having a specific ailment in this case) and credit card fraud detection are two other examples of binary categorization.

The following are well-known binary classification algorithms: logistic regression, K-nearest neighbors, decision trees, support vector machine, naive Bayes, etc.

Some algorithms, such as logistic regression and support vector machines, were created expressly for binary classification and do not by default support more than two classes.

8.2 Logistic Regression

One of the most often used machine learning algorithms, within the category of supervised learning, is logistic regression. Using a predetermined set of independent factors, it is used to predict the categorical dependent variable.

In a categorical dependent variable, the output is predicted via logistic regression. As a result, the result must be a discrete or categorical value. Rather than providing the exact values of 0 and 1, it provides the probabilistic values that fall between 0 and 1. It can be either Yes or No, 0 or 1, true or false, etc.

With the exception of how they are applied, logistic regression and linear regression are very similar. While logistic regression is used to solve classification difficulties, linear regression is used to solve regression problems.

8.3 Support Vector Machine

One of the most well-liked supervised learning algorithms, support vector machine (SVM) is used to solve classification and regression problems. However, it is largely employed in machine learning classification issues.

The SVM algorithm's objective is to establish the best line or decision boundary that can divide n-dimensional space into classes, allowing us to quickly classify fresh data points in the future. A hyperplane is the name given to this optimal decision boundary.

SVM selects the extreme vectors and points that aid in the creation of the hyperplane. Support vectors, which are used to represent these extreme instances, form the basis for the SVM method.
SVM comes in two varieties as follows:

Linear SVM is used for data that can be divided into two classes using a single straight line. This type of data is called linearly separable data, and the classifier employed is known as a Linear SVM classifier. It comes under binary classifier.
Non-linear SVM is used for non-linearly separated data. If a dataset cannot be classified using a straight line, then it is considered non-linear data, and the classifier employed is referred to as a non-linear SVM classifier.

8.4 Estimating Class Probabilities

A scoring classifier that outputs probability vectors across classes, or a mapping, is referred to as a class probability estimator. Where is the likelihood that class Ci will be present in example x, and in this scenario, we use to represent the estimated probability of the positive class for example x. If there are only two classes, then the probability associated with one class is equal to 1 minus the likelihood of the other class. We typically do not have direct access to the true probabilities, much like with scoring classifiers (x).

The Brier score, which uses this concept of inaccuracy in probability estimations, is a common term in forecasting theory. Equation 8.1's factor 1/2 ensures that the squared error for each example is normalized between 0 and 1; the worst-case scenario is when the incorrect class is predicted with probability 1, which results in two incorrect "bits." This becomes a single phrase that solely refers to the positive class when there are two classes. Remember that a class probability estimator becomes a classifier if it is "categorical" (it assigns probability 1 to one class and probability 0 to the others), in which case mean squared error (MSE) becomes accuracy.

$$\text{Brier Score}(\vec{x},\vec{y}) = \frac{1}{N}\sum_{i=1}^{N}(\hat{p}(x_i) - y_i)^2 \tag{8.1}$$

True positive refers to the situation in which the model correctly foresees the patients as being positive (TP). True negative is the term used when the model correctly forecasts the patients' outcomes as being negative (TN). Some patients may receive incorrect diagnoses from the binary classifier. False negatives are errors that occur when an ill patient is misclassified as healthy due to a negative test result (FN). Similar to this, a false positive occurs when a healthy patient is misdiagnosed as having an illness due to a positive test result (FP).

A binary classifier can be assessed using the following criteria:

- True Positive (TP): The patient has an illness, as predicted by the model.
- False Positive (FP): The patient is in good health, despite the model's "diseased" prediction.
- True Negative (TN): The patient is in good health, as predicted by the model.
- False Negative (FN): The model predicts "healthy" even when the patient is ill.

Accuracy Score of Binary Classifier = (TP + TN)/(TP + FP + TN + FN)

8.5 Confusion Matrix

A Figure 8.2 called a confusion matrix is used to describe how well a classification system performs. The output of a classification algorithm is shown and summarized in a confusion matrix. Examining the confusion matrix is a significantly more effective technique to judge a classifier's performance. The basic idea is to record the frequency with which occurrences of class A are categorized as class B. A confusion matrix has rows for each actual class and columns for each predicted class.

$$\text{Precision} = (\text{TP})/(\text{TP} + \text{FP})$$

Making one positive prediction and ensuring that it is accurate is a simple way to achieve perfect precision (precision = 1/1 = 100%). Because the

Confusion Matrix			
		Predicted	
		Positive	Negative
Actual	Positive	True Positive	True Positive
	Negative	False Positive	True Negative

Figure 8.2 Confusion matrix.

classifier would ignore all but one positive instance, this would not be very helpful.

$$\text{Recall} = (\text{TP})/(\text{TP} + \text{FN})$$

F1 Score:
By calculating the harmonic mean of a classifier's precision and recall, the F1-score integrates both into a single metric. It mainly used to compare the effectiveness of two classifiers. Assume that classifiers A and B have higher recall and precision, respectively. The F1-scores for both classifiers in this situation can be used to assess which one yields superior results.

$$\text{F1 Score} = 2(P * R)/(P + R)$$

where P = precession, and
R = recall of the classification model.

8.6 Beyond Binary Classification

Some ideas are, at their core, binary. For instance, it is difficult to generalize the idea of a coverage curve to more than two classes. Now, let us talk about some general problems with class probability estimation, scoring, and classification with more than two classes. Two topics will be covered in the discussion: how to assess multi-class performance and how to construct multi-class models from binary models. For some models, like linear classifiers, which are primarily meant to distinguish between two classes, the latter is essential. Other models, such as decision trees, automatically handle any number of classes.

When learning about classification issues in machine learning, it is common to start with binary classification, or situations where there are only two possible outcomes, like spam or not spam, fraud or not fraud, and so forth. Beyond that, we have **multi-class** and **multi-label** classification. Let us begin by outlining each of them.

8.7 Multi-Class Classification

Multiple class classification tasks are quite common. For instance, the doctor may want to further categorize a patient into one of numerous variations of rheumatic disease once the patient has been diagnosed with it.

The effectiveness of a classifier can be evaluated using a k-by-k contingency table if there are k classifications. If we are interested in the classifier's accuracy, which is still equal to the sum of the descending diagonal of the contingency table divided by the quantity of test instances, evaluating performance is simple. As before, this can mask differences in performance across classes, and other metrics might be more significant.

When your target variable contains more than two categories, you use multi-class classification (y). For instance, you might have a size scale of small, medium, large, and extra large (xlarge), or you might have a star rating system. These levels can also be thought of as classes. This strategy's goal is to predict one class out of the possible classes.

Instructing the classifier to handle more than two classes requires a strategy, which may have an impact on performance in terms of generalization or computational resources. As the reverse of overfitting, generalization refers to how well the classifier performs on future data that has not yet been observed.

When using the one-vs-rest (OVR), also known as the one-vs-all (OVA), technique, a single classifier is fitted for each class and tested against every other class. In essence, OVR divides the multi-class problem into a collection of binary classification issues. Because there are n classifiers for n classes, OVR typically performs well. The most widely used tactic is OVR, so you can begin your adventure here. This basically looks like this in practice:

- Classifier 1: small vs. (medium, large, and xlarge)
- Classifier 2: medium vs. (small, large, and xlarge)
- Classifier 3: large vs. (small, medium, and xlarge)
- Classifier 4: xlarge vs. (small, medium, and large)

If you have an extremely high number of classes, OVR typically does not scale well. One-on-one may be preferable. Next, let us discuss it.

For each pair of classes, the one-vs-one (OVO) method fits a single classifier. Here is how that appears:

- Classifier 1: small vs. medium
- Classifier 2: small vs. large
- Classifier 3: small vs. xlarge
- Classifier 4: medium vs. large
- Classifier 5: medium vs. xlarge
- Classifier 6: large vs. xlarge

Although the compute complexity is greater than the OVR strategy, it has the advantage that each classifier is fitted on a smaller subset of the data, as opposed to the OVR strategy, which fits each classifier on the entire dataset.

8.8 Multi-Label Classification

The supervised learning problem known as "multi-label classification" allows instances to be connected to multiple labels. This is a development of single-label classification (also known as binary or multi-class), in which each instance is only connected to a single class label.

There is a MultiOutputClassifier for multi-label categorization. This fits a classifier for each class, much like OVR. However, if appropriate, then this can produce numerous classes for a single prediction as opposed to a single predicted outcome.

In multi-label classification, the task is to predict the label sets of unseen instances by analysing training instances with known label sets. The training set is made up of instances, each of which is associated with a set of labels.

The difference between multi-class classification and multi-label classification is that, whereas the classes in multi-class problems are mutually exclusive, the tasks in multi-label problems are all related in some way.

A fruit can be either an apple or a pear, but not both at once, under the assumption that each sample is given one and only one label in multi-class classification, for instance. A text may simultaneously be about religion, politics, finance, or education, or it may not be about any of these things, which is an example of multi-label classification.

Predictive modeling's classification challenge entails producing a class label from an input. It differs from regression exercises where a numerical value is predicted. A classification task often entails foretelling a single label. Another option would be to forecast the likelihood over two or more class labels. The classification task presupposes that the input only belongs to one class in these situations since the classes are mutually exclusive. Predicting more than one class label is necessary in some classification tasks. This implies that belonging to a class or using a class label are not mutually exclusive. Multiple label classification, or multi-label classification for short, is the term used to describe these activities.

For each input sample in multi-label classification, zero or more labels must be produced simultaneously, and they must be needed for each output.

Reference

1. Ullrich, C., Descriptive and Prescriptive Learning Theories, in: *Pedagogically Founded Courseware Generation for Web-Based Learning*. Lecture Notes in Computer Science, vol. 5260, Springer, Berlin, Heidelberg, 2008, https://doi.org/10.1007/978-3-540-88215-2_3.

9
Model Selection

Model selection is the process of selecting the final learning machine model among a set of training data model training models.

Version choice is a technique that can be implemented to each exclusive forms of fashions [e.g., retrieval, support vector machine (SVM), and k-nearest neighbor (KNN)] and all models of the same type configured by different model parameters (e.g., one of a kind kernels in SVM).

If we have a variety of models of different complexities (e.g., line or regression models with different polynomials, or KNN separators with different K values), how should we choose the right one?

For example, we may have a database that has an interest in creating a segment or model that predicts retrospective. We do not know in advance which model will perform better on this issue, as it is unknown. Therefore, we measure and evaluate the list of different models in the problem.

Selecting models is the process of selecting one of the models as the final model that deals with the problem. Selecting models is different from model testing.

For example, we test or evaluate candidate models to choose the best one, and this is a choice of models. Although once the model is selected, it can be tested to determine how well it is expected to work normally; this is a model test.

The process of evaluating the performance of a model is known as model testing, whereas the process of selecting the appropriate level of model flexibility is known as model choice.

9.1 Model Selection Considerations

The simulation models are truthful, despite the fact that selecting among them is an actual assignment for the learning system used.

First, we need to get past the idea of the "quality" version.

Kolla Bhanu Prakash. Machine Learning for Industrial Applications, (127–148) © 2024 Scrivener Publishing LLC

All models have a few predictive mistakes, in terms of statistical noise inside the records, data sample imperfections, and obstacles for each form of unique model. Consequently, the concept of a great or ideal version does now not assist. As a substitute, we ought to search for a model "accurate enough."

9.1.1 What Do We Care Approximately When Choosing the Final Version?

Assignment members might also have specific necessities, consisting of upkeep and confined modeling complexity. Accordingly, a model with a low ability however easy and easy to recognize can be selected.

Alternatively, if the version's capabilities are valued chiefly other concerns, then the model's potential to carry out nicely outside the pattern can be decided on no matter the gadget complexity involved.

Therefore, an extremely good enough version may also discuss with many things and be precise to your undertaking, which include the following:

- A version that meets the desires and demanding situations state-of-the-art task participants.
- Model capable enough while thinking about the time and assets available.
- Powerful version compared to non-smart fashions.
- Competitive version associated with other examined fashions.
- State of the art expertise-associated version.

Next, we need to recollect the choice. For instance, we do now not pick the suitable version, as all models may be discarded. This is due to the fact, as soon as we have selected a model, we are able to equal the remaining new model in all to be had information and begin using it to make predictions.

So, do we pick out most of the algorithms used to in shape the models in the training database?

Some algorithms require special information correction to appropriately expect problem formation inside the cutting-edge algorithm. Therefore, we must cross further and recall selecting fashions as a manner today's among the pipeline improvement fashions.

Each pipe may also take from the identical uncooked training database and produce a model that may be examined inside the same way but may also require exceptional or extra complex calculation measures, consisting of the following:

- Filtering facts.
- Information transformation.
- Feature selection.
- Engineering component.

And extra...

If you appearance intently at the task latest models, then you will locate many nuances.

Now that we have come to be acquainted with present day the concerns worried in choosing a model, allow us to evaluation some not unusual strategies for selecting a version.

9.2 Model Selection Strategies

The nice way to pick out a model calls for "sufficient" information, which may not be complete relying at the complexity trendy the trouble.

In this perfect case, we can cut up the records into education, validate, and take a look at set; then insert candidate models into the schooling set; test them and select them from the validation set; and record the very last model performance to the check set.

If we are in a facts-wealthy environment, the pleasant manner [...] is to randomly divide facts into three parts: education set, verification set, and check set. The schooling set is used to healthy the models; a verification set is used to estimate the mistake in version selection; the test set is used for widespread error checking ultra-modern the selected very last version.

This does not apply to many modeling issues given that we do now not commonly have enough records, or we will even decide what is not enough.

In many systems, but, the provision modern training and checking out data can be restricted, and, to construct good models, we want to use as an awful lot facts as feasible in schooling. But, if the verification set is in minimum, then it will provide an exceptionally correct measure cutting-edge the expected overall performance.

Instead, there are two primary categories today's strategies for measuring exact model selection; ultra-modern direction:

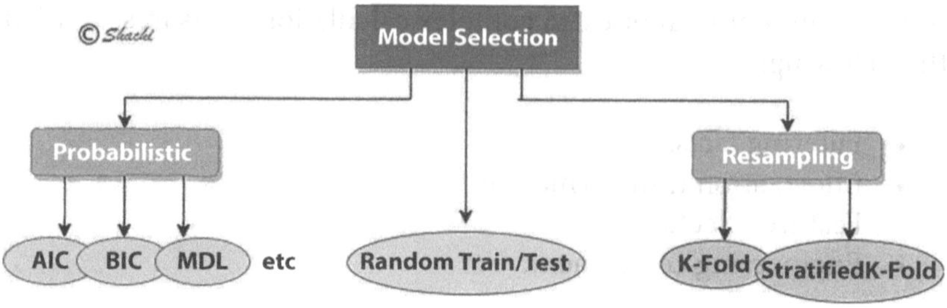

Figure 9.1 Model selection strategies. Source: https://miro.medium.com/max/1196/1*s2NqGsdIg2lXmiDx-cLlcg.png [1].

- Possible Steps: Pick out a model with an internal error sample and complexity [1].
- Re-Sampling Strategies: Choose a model with limited blunders outdoor the sample [1].

Let us examine one aspect at a time.

9.3 Types of Model Selection

9.3.1 Methods of Re-Sampling

Re-sampling strategies, because the name suggests, are easy approaches to arrange data samples for trying out if the version plays well on untrained records samples. In different words, re-sampling allows us to recognize if the model will produce nicely.

9.3.2 Random Separation

Random separation is used to randomly pattern a percent of information on education, trying out, and great warranty units. The gain of this approach is that there is a superb chance that the real populace is nicely represented in all three organizations. In extra formal phrases, random segregation will save you from biased sampling of facts.

It is very critical to note the usage of validation set within the model choice. The verification set is the second check set and one might ask, why did you have two test sets?

Within the system of choosing capabilities and modeling of models, a take-a-look-at-set is used for model checking out. This is the manner the model parameters and the characteristic sets are selected so that they provide the entire result in the check set. Consequently, a verification set with completely identifiable records elements (no longer used in tuning modules and deciding on features) is used for very last checking out.

9.3.3 Time Divide

There are other varieties of statistics wherein random splitting does no longer arise. As an instance, if we need to educate a weather forecasting version, we cannot randomly cut up data into schooling and trying out units. This may encompass the season pattern! Such records are regularly known as Time Line.

In such cases, a wise department of time is used. The schooling set may also comprise information for the closing 3 years and 10 months for the cutting-edge year. The final months can be reserved for a take a look at or verification set.

There may be moreover the idea of window gadgets, in which the version is educated up to a high quality date and tested inside the coming days repeatedly in order that the schooling window is always bendy over the identical day (as an end result, the check set also decreases via day). The benefit of this technique is that it strengthens the model and prevents overset up when the test set is simply too small (say, 3 to 7 days).

However, the regression of time series data is that activities or records factors are not together exceptional. One occasion may also have an effect on all the following information entries.

For instance, adjustments within the ruling party may additionally have dramatically changed the demographics of the subsequent years. Or the notorious coronavirus can have a massive effect on economic records for the following few years.

No gadget learning model can research from beyond statistics in such a state of affairs because the information pointing earlier than and after the occasion has great differences.

9.3.4 K-Fold Cross-Validation

The shortcut authentication method works through randomly shuffling the records and dividing it into ok agencies. Then, while repeating each group, the group wishes to be considered as a take a look at set at the same time as all the different agencies are grouped collectively right into a schooling set. The version is tested in the check institution and the manner keeps in companies okay.

Therefore, on the quit of the manner, one story distinct k results in special test companies. The excellent version can be effortlessly selected by means of selecting with high rankings

9.3.5 Stratified K-Fold

The stratified ok-Fold procedure is just like that of okay-Fold authentication with one factor of difference; in evaluation to the alternative ok-fold confirmation, the values of the goal variable are considered within the stratified okay-fold.

If, for example, the target variable is a two-magnificence segment variable, then the k-coiled curve guarantees that each test roll gets an equal ratio of training in comparison to a schooling set.

This makes model trying out greater correct and version training less biased.

9.3.6 Bootstrap

Bootstrap is one of the maximum powerful methods to find a stable version. It is a protracted way close to the technique of random separation as it follows the idea of random sampling.

The first step is to select a sample length (typically the size of the authentic database). Thereafter, the sample records factor must be randomly decided on from the real database and brought to the bootstrap pattern. After including, the sample needs to be back to the authentic sample. This manner needs to be repeated N times, where N is the pattern size.

Consequently, it is far a repetitive technique that creates a bootstrap pattern by using taking facts points from the original database via changing, which means that a bootstrap pattern can comprise multiple times of the identical facts point.

The version is educated in a bootstrap pattern and tested for all those facts factors that did not attain the bootstrap sample. Those are referred to as out-of-bag samples.

9.3.7 Possible Steps

The probabilistic measures do not forget not only the most effective general performance of the model but also the complexity of the version. Model complexity is a diploma of a model's potential to seize information versions.

For instance, a tremendously biased model along with linear regression set of rules is an awful lot much less complex, and, on the other hand, the neural community is a lot better in complexity.

Every other critical point to be aware right here is that the overall performance of the model considered inside the feasible steps is calculated most effective in the training set. A backup take a look at set is typically now not required.

A touch wrong, however, lies in the truth that the feasible measures do now not bear in mind the uncertainty of the fashions and have the possibility to choose fashions that are less difficult than complicated models.

9.3.8 Akaike Information Criterion (AIC)

It is far recognized that every version is completely faulty. There is usually a lack of statistics that may be measured the usage of the KL information metric. Kulback-Liebler or KL divergence measures the difference in the distribution of opportunities for two versions.

The mathematician, Hirotugu Akaike, taken into consideration the connection among KL know-how and higher opportunities (much more likely, one would love to growth the conditional possibilities of viewing information factor X, given the parameters and distribution of certain possibilities) and evolved the notion of records term (or IC). Therefore, AIC is a measure of records loss. That is how the distinction among the two extraordinary fashions is taken and the version with the least loss of information is recommended as the desired model [2].

$$\mathbf{AIC = (2X - 2\log(Y))/Z}$$

https://neptune.ai/blog/the-ultimate-guide-to-evaluation-and-selection-of-models-in-machine-learning [2].

X = no of independent variables or predictions;
Y = multiple model possibilities;
Z = no of data points in the training set (especially useful if there are small data sets).

The limit of the AIC is that it is not very good with standard models as it tends to choose more complex models that lose less training experience.

9.3.9 Bayesian Information Criterion (BIC)

BIC is derived from the concept of Bayesian possibilities and is appropriate for fashions skilled beneath the highest probability.

$$\mathbf{BIC = X^{*}log(Z) - 2log(Y)}$$

https://neptune.ai/blog/the-ultimate-guide-to-evaluation-and-selection-of-models-in-machine-learning [2].

X = no of independent variants;
Y = more opportunities;
Z = No of sample/data points in the training set.

BIC punishes the model for its complexity and is best used if the information length is not very small (otherwise, it tends to live on very simple models) [2].

9.3.10 Minimum Definition Length (MDL)

MDL is primarily based on data idea concerning values inclusive of entropy that measures the average number of bits had to constitute an event from viable distribution or random variables.

MDL or minimal definition period is the minimal quantity of such bits required to represent the version [2].

$$\mathbf{MDL = L(m) + L(D|m)}$$

https://neptune.ai/blog/the-ultimate-guide-to-evaluation-and-selection-of-models-in-machine-learning [2].

d = model;
D = model predictions;
L (m) = no of bits required to represent the model;
L (D | m) = no of bits needed to represent the predictions from the model.

9.3.11 Building Risk Reduction (SRM)

Device mastering models face the inevitable undertaking of defining commonplace theory from a limited set of records. This ends in overcrowding situations in which the version takes sides within the schooling information that is its most important supply of studying. SRM seeks to measure the complexity of the version in opposition to its fulfillment in facts capture [2].

9.3.12 Excessive Installation (Overfitting)

If we have a large set of models in the model class in relation to the amount of data, it is possible that the estimated model performs well on the data used to run the parameters in it, but that this model performs improperly negatively charges.

Example of AR model with white noise u:

$$\boldsymbol{x(t) + a_1 x(p-1) + \ldots\ a_p x(p-t) = u(p)}$$

- true AR model has order $t = 5$;
- the parameters to be imagined are $\theta = (a_1, a_2, \ldots, a_t)$ with t as the unknown;
- define a quadratic loss function

$$f(y\theta) = \sum_{p=t+1}^{N} |\mathbf{x(p) - a1x(p-1) + \cdots + atx(p-t))|2}$$

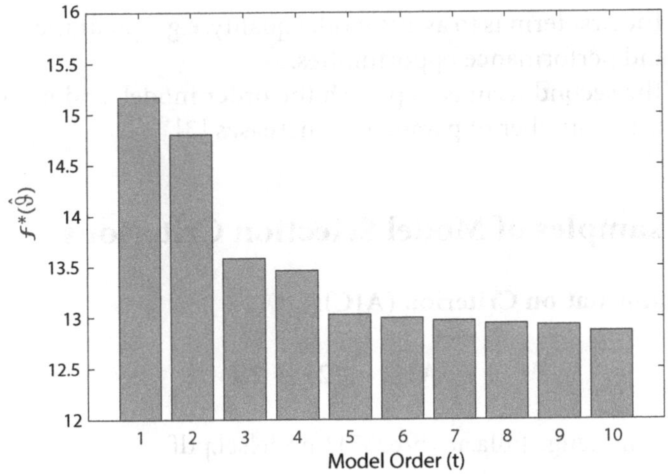

Figure 9.2 Overfitting of model.

and obtain $\hat{\theta}$ by using the LS method:

$$\hat{\theta} = (\hat{a}_1, \hat{a}_2, \ldots, \hat{a}_t) = \mathbf{argminf}(\theta)$$

- The decreasing function of the model structure is represented by the minimized loss.
- *f* begins to drop as the model pulls up the relevant methods.
- The version tends to over suit the information because the t will increase.
- In widespread, we look ahead for the "knee" in the graph (around $t = 5$).

9.4 The Principle of Parsimony

Aim: Among all competing models that define the data well, a model with a small number of parameters should be selected.

In the previous example, we saw how to determine model order *t*?

- Trading curve between job loss and model planning
- Model selection criteria

The condition of choosing a model has two parts: **loss function + model complexity**

- The first term is to assess model quality, e.g., quadratic losses and performance opportunities.
- The second term is to punish the order model, and it grows as the number of parameters increases [3].

9.5 Examples of Model Selection Criterions

Akaike Information Criterion (AIC):

$$\mathbf{AIC = -2L + 2}\boldsymbol{d}$$

http://jitkomut.eng.chula.ac.th/ee531/modelsel.pdf

Bayesian Information Criterion (BIC):

$$\mathbf{BIC = -2L} + \boldsymbol{d}\mathbf{log}\boldsymbol{N}$$

http://jitkomut.eng.chula.ac.th/ee531/modelsel.pdf

Akaike's Final Prediction-Error Criterion (FPE):

$$\mathbf{FPE} = \frac{\mathbf{1}}{\boldsymbol{N}} \sum_{t=1}^{N} e^2(t,\vartheta)\,!\,\frac{\mathbf{1} + d/N}{\mathbf{1} - d/N}$$

Source: http://jitkomut.eng.chula.ac.th/ee531/modelsel.pdf.

- This is a feasible task;
- d number of active parameters;
- E (t, θ) predictive error.

9.6 Other Popular Properties

- BIC usually punishes more complex models (due to the term log N).
- BIC has not changed (probability that BIC will select the correct model is closer to one such as sample size N → ∞).
- AIC and FPE usually select more complex models such as N → ∞ [3].

The most pragmatic method is to use the verification method as follows. Consider a collection of limited models in the structures of different models. For example, adjust the model type, and create order parameters for different model models. Then, you can find each candidate using the appropriate validation clause. The model structure (order) that leads to the best points is clearly chosen. This method is often used for design parameters without direct physical translation.

9.7 Key Considerations

Selecting models is simply a validation of the modeling experience. One can easily assume that a person with experience in the field does not need to use any of the above methods to guide his or her design decisions. But to help such an expert, trading strategies are important, a few of which are listed here.

- Organize data in different ways.
- View trends or seasonal results in data.
- Is the LTI model sufficient or our requirements?
- View data spectra.
- Try recording and explaining the audio process. Where does the sound come from in your study? So, what
- Try a simple model that does not make sense, and then try to figure out how you can determine if this is not enough for you. Also, if you can formalize the specific goal you are working on, you are in the middle of the modeling process.

9.8 Model Validation

In machine learning, model validation ensures that models provide satisfactory results in their input data, with both quality and value objectives. Although, in part, it includes a set of tried and true methods, model validation is a separate process that can be easily compressed or marked and applied to all models, creating opportunities for creativity and ingenuity.

Model validation is part of controlling machine learning, the whole process of how an organization manages access, applies policy, and tracks the work of models.

9.8.1 Why is Model Validation Important?

Model verification ensures the efficiency and accuracy of the trained model in preparation for use. Without version validation, the version would not feature well and training time will no longer be to be had. A poorly validated model may not be strong enough to adapt to new stress conditions or it may be overused to adopt and use new inputs properly. Model validation will take place before the model is installed on a complete website. Monitoring will always occur near the active model.

9.8.2 How to Validate the Model

There are unique ways to confirm model statistics: one can test a model on a trained data model, or one can test it on an external test set. The first method introduces the problem of overloading: one can incorrectly install any database at the expense of creating an inaccessible model for additional data. If the model is made complete on a single website, then the model may not be able to use and identify relevant results with new data and thus will not be valid. One can take the form of trying to make a pair with a pair (a, b) when given 100 of them as a training set.

High-level polynomial may fit data into a training set exactly, whereas it is more sensitive to data outside the training set. It is a common practice instead of validating models using a test set. Once given a data set, one can create a random set by extracting 10% to 20% of the data. In the case of 100 pairs (a, b), one can find that a high-level polynomial is easily entered by separating the test set and the test model against it. One may choose to use a simpler model, such as a line cut, which has a higher chance of exceeding model validation.

Many different mathematical test metrics can be used to validate the model in general, including the average measurement error, the error square, and the ROC curve [4].

9.8.3 What is a Model Validation Test?

Model validation testing is a process of testing the effectiveness of models compared to real data. It is important that the model is verified by considering the features and components before introducing them to the production ecosystem.

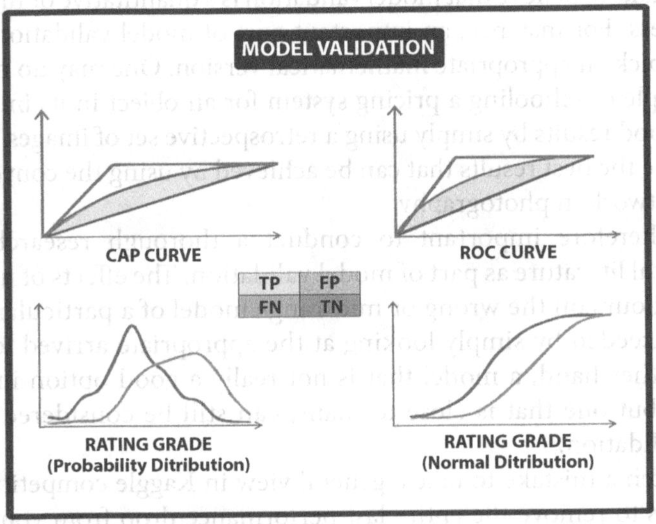

Figure 9.3 Model validation. Source: http://test.basel.in/wp-content/uploads/2019/09/MODEL-VALIDATION-1.jpg [4].

- Validation: The process of building the right model.
- Verification: The process of model development in the right way.
- Measurement: The loop process of comparing the model to a real system and making adjustments accordingly.

9.8.4 Benefits of Modeling Validation

Initial Detection of Deficiencies and Errors: With the model-backed test, it is easy to detect errors and errors before model verification. Cost Reduction: It is easy to reduce the different types of costs as the model validation test analyzes errors so that they are not ignored. Finding More Defects and Errors: When a disability is ignored, it is recognized very quickly. Model Exceptional Development: Improvement is a post process for detecting extra shortcomings and mistakes. Providing Stability and Flexibility in the Software Development Process: With the proper use of the model verification test, the scope of flexibility and flexibility is delivered to the process. Data Analysis and Model Related Information: Data are the backbone of the model and, with the help of model testing proper interpretation of data and information, can be done.

9.8.5 Model Validation Traps

It is a mistake to believe that model validation is a quantitative or mathematical process. For instance, an important part of model validation ensures that you pick an appropriate mathematical version. One may do not forget the example of schooling a pricing system for an object in its image. One can get good results by simply using a retrospective set of images. But, this will ignore the best results that can be achieved by using the convolutional neural network in photography.

It is therefore important to conduct a thorough research of the mechanical literature as part of model validation. The effects of unlimited working hours on the wrong or mid-range model of a particular website can be exceeded by simply looking at the appropriate arrived locations. On the other hand, a model that is not really a good option in a given data set, but one that is close to many, can still be considered without model validation.

It is often a mistake to take a general view in Kaggle competitions that the goal is to remove the entire last performance drop from your model.

Repeating the problem of machine learning with a different model carries the problems of expensive digging, time consuming, and erroneous tendencies. It is miles regularly genuine that there's only one "right" version at the internet site, as is the case with large records units and sensor networks. Or that there is no "right" model, and several will come close to good, as is the case with many Kaggle competitions that are not based on the image.

9.8.6 Data Verification

Data verification is another important part of model validation. Data values can be corrupted or contain errors in ways that damage the results of model training. The integrity of data values can be verified by means of self-examination in data categories, systematic analysis, or graphs. Integrity can also be properly assessed by ensuring that the data are taken from a reliable, reliable, well-maintained, and up-to-date source.

Training data and test sets should be taken from the distribution of the same opportunities or closer to achieving adequate results. In addition, there is a risk that models may be at risk for errors in certain input data values because they are poorly represented in the training set. In the event that such a category of errors is possible, it is important to ensure that the training set properly covers all of the data inputs where the model will need to be tested, or you may lose model validation. There are many ways to ensure that a training set is adequate, which includes searching for data or building up its visual episodes.

It is very important that we make accurate assumptions about the similarities between the training set and the model data that will eventually be tested, which is also a quality process.

9.8.7 Model Performance and Validation

Setting standards for how a model should work is another part of modeling and validating machine learning. Models will never be completely accurate, and a change-off has to be made between threat of errors and education time and the size of the training set. Ultimately, the decision should be made with quality, perhaps through the process of testing many different models in the database to determine the standards. Sometimes, no model works well enough and the project has to be completely scrapped. One might consider the issue of self-driving cars, a very serious problem for which there is no real solution.

9.9 Self-Driving Cars

Because self-driving cars are already found to be operating to a limited extent in testing settings, the right balance between success and failure is key to figuring out whether self-riding motors may be a significant 21st century technological success or a main source of illicit investment. Self-driving cars also provide an example of a case where the line between success and failure of a model cannot be reduced to number one. Merequins per mile, deaths per mile, errors per mile, number and type of traffic violations per mile, and number of individual interventions per mile can be important metrics for model validation, and planning through them to determine between success and failure can be a complex process, with both quality and value.

9.10 K-Fold Cross Validation

There are many ways to compare models, which is an important part of model validation. K-fold cross validation is one such method. To ensure k-fold contrast, the training set is randomly divided into k-groups. Then, with repeated k times, one person holds each unique group, trains the model to the remaining data, and evaluates the model in that group. Performance is estimated at each k for the opposite verification.

One can use k-fold cross validation to compare and select models before testing in a complete test set. The Akaike information concept is another way of comparing models. The AIC essentially measures the quantity of records misplaced with the aid of a specific version. AIC is given as

$$\mathbf{AIC = 2k - 2ln\ (L^\wedge)}$$

where L^ is the highest possible modeling function. Because logarithms grow monotonically, as well as ln (L^) term is deleted, the preferred models are those with a lower AIC value. AIC may be helpful, for example, in validating a model if you want to compare the performance of a less efficient (but faster training) model to a more efficient (but slower training) model. If the inefficient model is close enough to the AIC to the most efficient model, then it may go beyond model validation.

9.11 No One-Size-Fits-All Model Validation

Ultimately, model validation can be a very traditional and very different process depending on the model and the database being played. There is no global system, process, or model validation method that is ready to validate any given model in any given database. As an example of the extent to which model validation methods may vary, one might consider the matter of neural layer image separation. Recent research has shown that the adversarial neural network can be trained to produce data inputs where the image filter can reveal large statistical errors.

Neural networks are considered to be the most vulnerable to such enemy attacks, so if the errors generated or the amount of data from which errors can be generated are considered too large, one can switch to a model that does not include neural networks to pass model validation. This model has the actual use of a self-driving system, which the neural network may determine, based on image input, speed, and driving mode. It would be extremely important for such a system to be able to withstand the onslaught of enemies, as criminals can place images on or near the road that often cause the motorists to make poor decisions. In general, training conflicting models against your model and evaluating results is a promising way to validate the model.

9.12 Validation Strategies

The validation step is wherein we can increase the hyperparameters for every set of rules we need to use. With the intention to carry out this verification step, we can use 80% of the data we created all through the break up test check step.

There are two most common verification strategies: K-roll verification and a hold-out set [5].

9.13 K-Fold Cross-Validation

To ensure the difference in K, we take our training set and divide it into classes or wrap "ok," in which "k" represents a certain quantity, inclusive of 3.K-fold works by shoving each wrap, so that each wrap gets the possibility of having both a training set and a confirmation set [5].

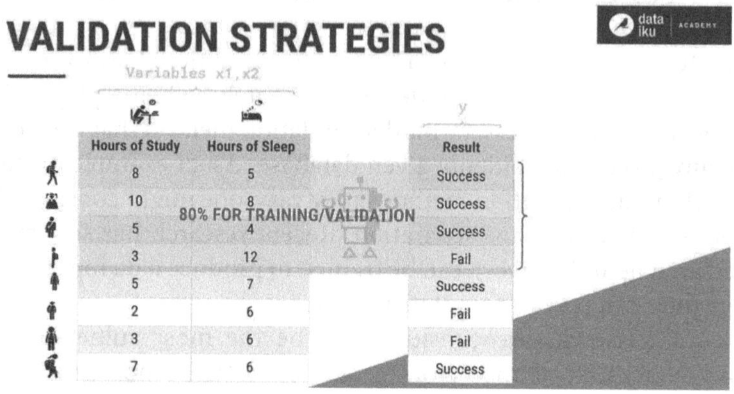

Figure 9.4 Validation strategies. Source: https://knowledge.dataiku.com/latest/courses/intro-to-ml/predictive-modeling/model-validation-summary.html.

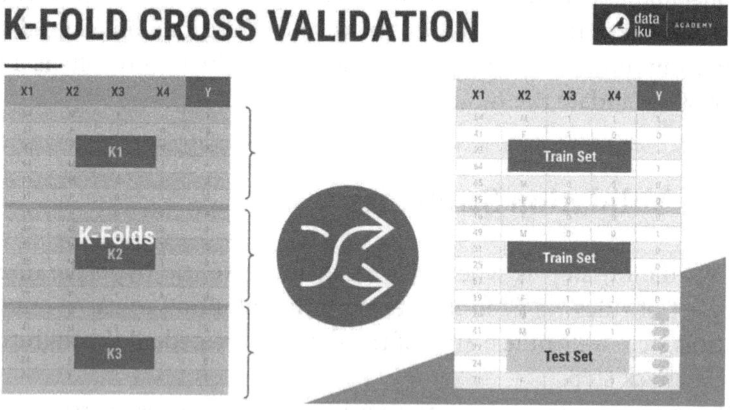

Figure 9.5 Training set and testing set split it into "K" sections or folds. https://knowledge.dataiku.com/latest/courses/intro-to-ml/predictive-modeling/model-validation-summary.html.

For every combination of hyperparameter values, we educate the version within the folders assigned to the schooling, and take a look at and calculate the error within the curve supplied for verification. The folds are then swiped in a spherical-robin fashion, until the error is calculated on all K-folds [5].

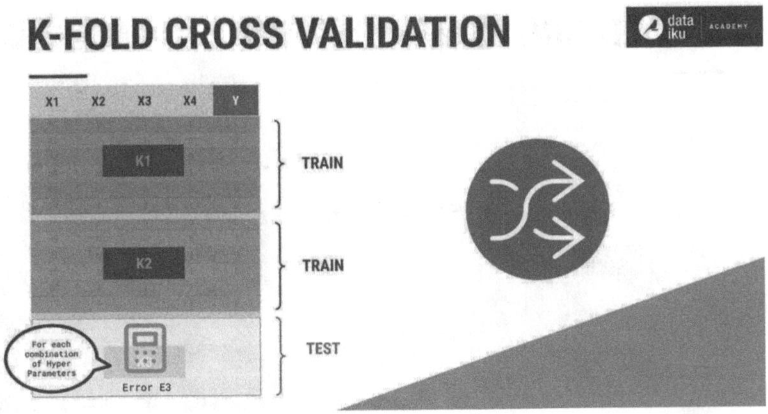

Figure 9.6 Error calculation on the fold assigned for validation. Source: https://knowledge.dataiku.com/latest/courses/intro-to-ml/predictive-modeling/model-validation-summary.html.

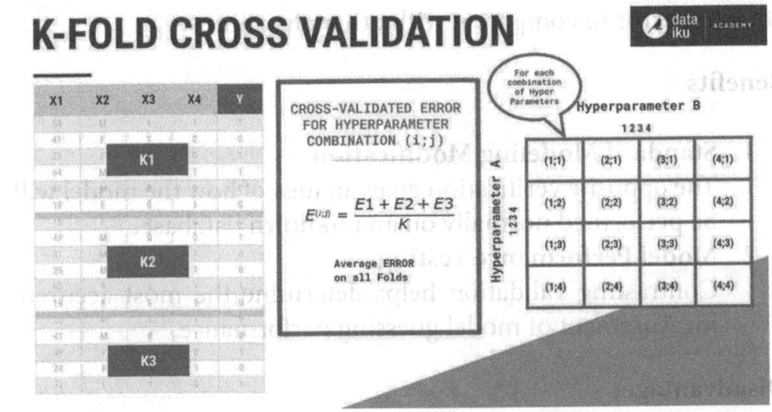

Figure 9.7 Cross-validated error for each combination of hyperparameter. Source: https://knowledge.dataiku.com/latest/courses/intro-to-ml/predictive-modeling/model-validation-summary.html.

The magnitude of these errors is our relatively verified error in each hyperparameter combination.

We then selected a combination of hyperparameters with a validated error, trained our model with a complete training set, and calculated the

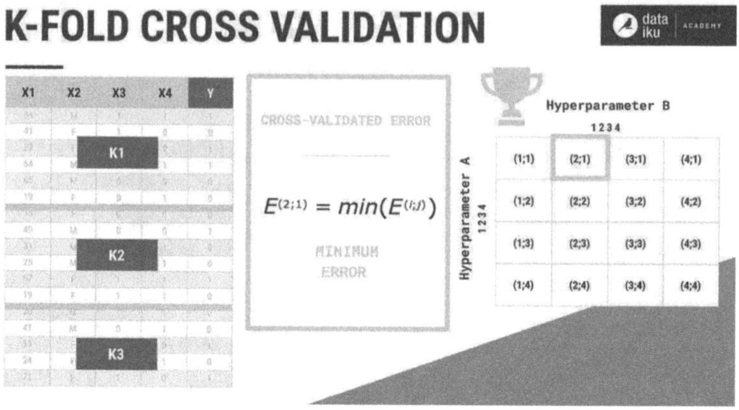

Figure 9.8 Choosing the combination having the best cross-validated error. Source: https://knowledge.dataiku.com/latest/courses/intro-to-ml/predictive-modeling/model-validation-summary.html [5].

error in the test set—the ones we have not touched still now. We can then use this test error to compare it with other algorithms.

Benefits

1. **Standard Modeling Modification**
   The opposite verification gives an idea of how the model will be performed normally on an unknown database.
2. **Model Performance Testing**
   Contrasting validation helps determine the most accurate measurement of model guessing performance.

Disadvantages

1. **Advanced Training Time**
   With the opposite assurance, we need to train the model in multiple training sets.
2. **Expensive Calculation**
   Conflict verification is very expensive as we need to train with multiple training sets.

9.14 Capture Confirmation Using Hold-Out Validation

Every other strategy is referred to as hold-out validation. On this approach, we really set aside part of our schooling set and use it as an affirmation set. As an example, we ought to create a 60%-20%-20% cut up (educate validation take a look at). On this strategy, the units do now not push [5].

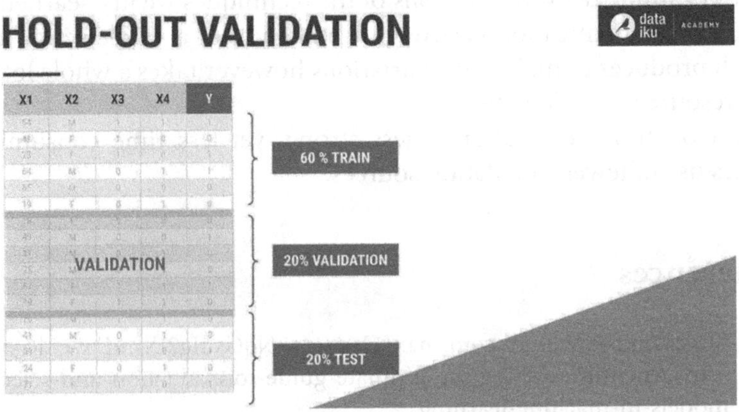

Figure 9.9 Hold-out validation. Source: https://knowledge.dataiku.com/latest/courses/intro-to-ml/predictive-modeling/model-validation-summary.html.

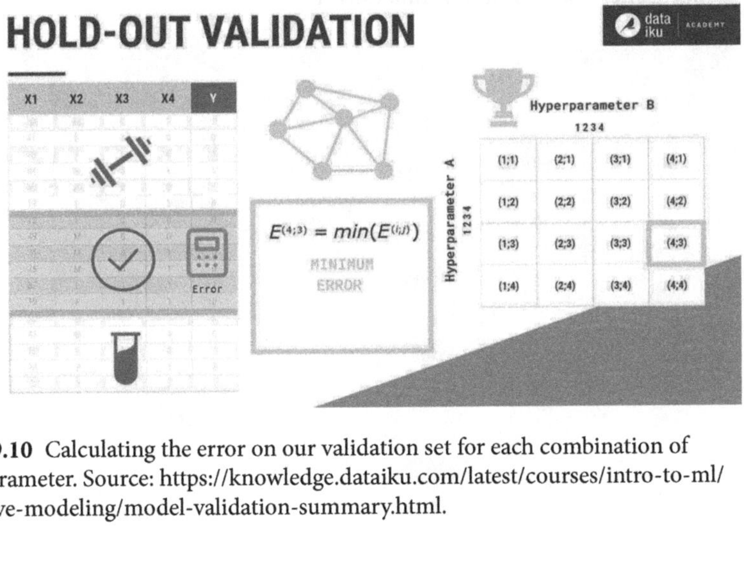

Figure 9.10 Calculating the error on our validation set for each combination of hyperparameter. Source: https://knowledge.dataiku.com/latest/courses/intro-to-ml/predictive-modeling/model-validation-summary.html.

As a substitute, we will calculate the mistake in our verification set for each aggregate of hyperparameter values that we want to test. Then, we pick out the bottom mistakes version and calculate the test error in our check set. This offers us a model that we can hopefully use for brand spanking new, invisible information [5].

9.15 Comparison of Validation Strategy

Let us examine the pros and cons of the techniques we just learned.

K-fold authentication is strong. It make use of a whole set of training, which produces a small visual variations however takes a whole lot of time and resources to calculate.

Not on time verification is less strong, yet less time consuming and makes use of fewer calculation sources.

References

1. https://miro.medium.com/max/1196/1*s2NqGsdIg2lXmiDx-cLlcg.png
2. https://neptune.ai/blog/the-ultimate-guide-to-evaluation-and-selection-of-models-in-machine-learning
3. http://jitkomut.eng.chula.ac.th/ee531/modelsel.pdf
4. http://test.basel.in/wp-content/uploads/2019/09/MODEL-VALIDATION-1.jpg
5. https://knowledge.dataiku.com/latest/courses/intro-to-ml/predictive-modeling/model-validation-summary.html

10

Support Vector Machines

Support vector machine or SVM is one of the greatest popular supervised learning algorithms concentrating on calculations, which is utilized for arrangement notwithstanding regression issues. However, normally, it is miles utilized for class issues in gadget getting to know.

The place of the SVM computation is to settle on the pleasant line or choice cutoff that can confine n-layered locale into getting ready with the objective that we can, without much of a stretch, put the new real factor points in the proper class in the fate. This most ideal choice cutoff is insinuated as a hyperplane.

SVM picks the over the top centers/vectors that help with turning into the hyperplane. These outrageous events are suggested as help vectors and, as needed, are set of rules that are named as assistance vectors with machining. Consider the under diagram where two exceptional classes are organized the utilization of a decision limit or hyperplane.

10.1 History

The principal SVM set of rules was envisioned by Vladimir N. Vapnik and Alexey Ya. Chervonenkis in 1963. In 1992, Bernhard Boser, Isabelle Guyon, and Vladimir Vapnik prescribed a method for making nonlinear classifiers by using the piece trick to most-edge hyperplanes. The fragile "edge" appearance, as is, by and large, applied in programming program applications, becomes proposed through Corinna Cortes and Vapnik in 1993 and distributed in 1995 [3].

10.2 Model

SVM can be understood with the example that we have used in the KNN classifier. Suppose we see a strange cat that also has some features of dogs,

Kolla Bhanu Prakash. Machine Learning for Industrial Applications, (149–176) © 2024 Scrivener Publishing LLC

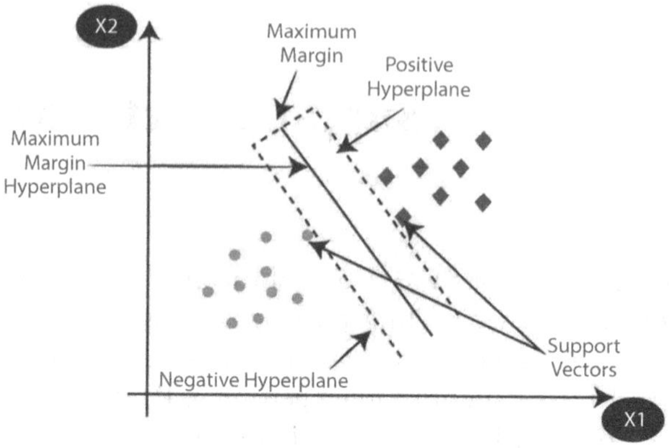

Figure 10.1 Support vector machine.

so if we want a model that can accurately identify whether it is a cat or dog, so such a model can be created by using the SVM algorithm. We will first train our model with lots of images of cats and dogs so that it can learn about different features of cats and dogs, and then we test it with this strange creature. So as support vector creates a decision boundary between these two data (cat and dog) and choose extreme cases (support vectors), it will see the extreme case of cat and dog. On the basis of the support vectors, it will classify it as a cat. Consider the under chart.

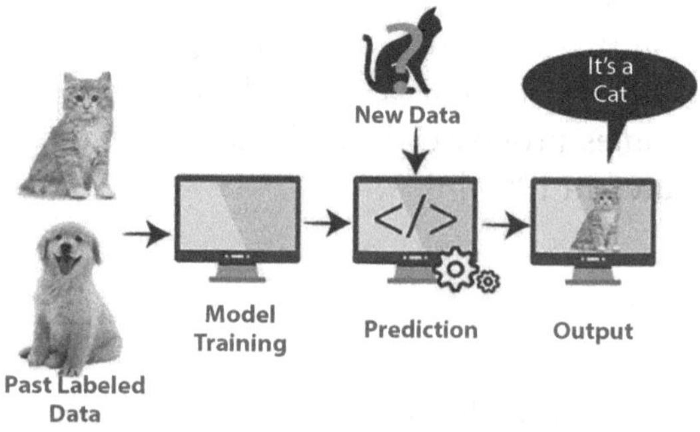

Figure 10.2 Model building and prediction.

SVM estimation may be used for the following:

- Automatic face recognition,
- Picture characterization,
- Text arrangement, etc.

10.3 Kinds of Support Vector Machine

SVM is of two sorts:

1. Straight Support Vector Machine
2. Non-Direct Support Vector Machine

10.3.1 Straight SVM

Direct SVM is utilized for straightly distinct records, which implies that, if a dataset might be arranged into illustrations through utilizing a solitary quickly line, then, at that point, such measurements are named as straightly divisible information; furthermore, classifier is used known as linear SVM.

10.3.2 Non-Direct SVM

Non-linear SVM is used for non-straightly confined real factors, which suggests that, if a dataset cannot be set up through the use of a promptly line, then, at that point, such records are named as non-direct data, and classifier utilized is inferred as non-direct SVM [1].

10.3.3 Benefits of Help Vector Machines

- Powerful in high layered regions.
- By the by strong in cases where a few aspects are more noteworthy than the amount of tests.
- Utilizes a subset of preparing factors inside the choice component (called help vectors), so it is likewise memory green.
- Adaptable: Remarkable kernel capacities might be focused on for the choice trademark. Ordinary pieces are outfitted, but it is additionally achievable to indicate custom kernels [2].

10.3.4 The Negative Marks of Help Vector Machines

- Assuming that the amount of elements is tons more than the wide assortment of tests, keeping off over-becoming in settling on kernel elements and regularization time-frame is fundamental.
- SVMs do not immediately offer freedom assesses, these are determined utilizing a profoundly valued 5-overlap move-approval (see appraisals and possibilities, under) [2].

10.3.5 Applications

SVMs can be utilized to determine different genuine worldwide issues:

- SVMs are important in text and hypertext request, as their utility can, on a very basic level, decrease the requirement for organized coaching models in each the norm and can destroy inductive and transductive settings. A few systems for shallow semantic parsing rely totally upon guide vector machines.
- Grouping of sneak peaks can, moreover, be finished using SVMs. Preliminary results show that SVMs acquire altogether favored request accuracy over customary inquiry refinement plans after simply three to four rounds of significance remarks. That is, besides being good "old" fashioned for picture division structures, close by those utilizations of an adjusted model of SVM that incorporates the really inclined directions of procedure as recommended through Vapnik.
- Class of satellite records like SAR information using managed SVM.
- A few techniques for shallow semantic parsing rely totally upon manual vector machines.
- The SVM set of rules has been thoroughly executed in regular and different sciences. They were used to sort proteins with up to 90% of the combinations organized capably. Change assessments on a very basic level ward on SVM loads have been recommended as an instrument for the understanding

of SVM models. Guide-vector system loads have also been used to translate SVM styles beforehand [3].

10.4 Hyperplane and Support Vectors Inside the SVM Set of Rules

10.4.1 Hyperplane

There are different lines/choice deterrents to confine the classes in n-layered space; however, we really want to observe the great choice breaking point that helps with requesting the records factors. This charming cutoff is insinuated as the hyperplane of SVM.

The components of the hyperplane rely upon the components present inside the dataset, which suggests that expecting there are two limits; then, the hyperplane can be a rapid line. Other than that, if there were three cutoff points, then, at that point, the hyperplane can be a two-size plane.

We ceaselessly make a hyperplane that has a greatest edge, on account of this the most distance among the records habitats.

10.5 Support Vectors

The estimation centers or vectors that can be the closest to the hyperplane and that influence the circumstance of the hyperplane are named as help vector. Considering that those vectors help the hyperplane, in this manner, it is known as an assistance vector [1].

10.6 SVM Kernel

The SVM portion is a capacity that takes low layered information region also transforms it into better-layered area, i.e., it changes not unmistakable issue over to separable issue. It is far ordinarily valuable in non-straight detachment issues. The truly positioned portion does some incredibly complicated data changes and then, at that point, discovers the cycle to divide the insights based absolutely at the names or results characterized [4].

10.7 How Can It Function?

As of now, the duplicating question is "How are we prepared to track down the authentic hyperplane?" Unwind, it is, by and by, not for the most part so particularly inconvenient as you would presume!

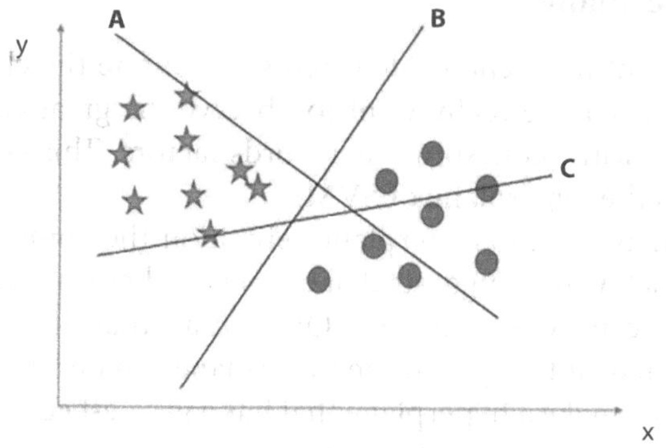

Figure 10.3 Circumstance 1.

10.7.1 See the Right Hyperplane (Circumstance 1)

Here, we have three hyperplanes (A, B, and C). By and by, pick the right hyperplane to organize stars and circles.

You truly need to recall a thumb rule to pick the fitting hyperplane: "select the hyperplane which separates the 2 headings better." On the present circumstance, hyperplane "B" has incredibly finished along these lines.

10.7.2 See the Appropriate Hyperplane (Situation 2)

Here, we have three hyperplanes (A, B, and C), and all are disconnecting the arrangement fittingly. As of now, how are we prepared to see the right hyperplane?

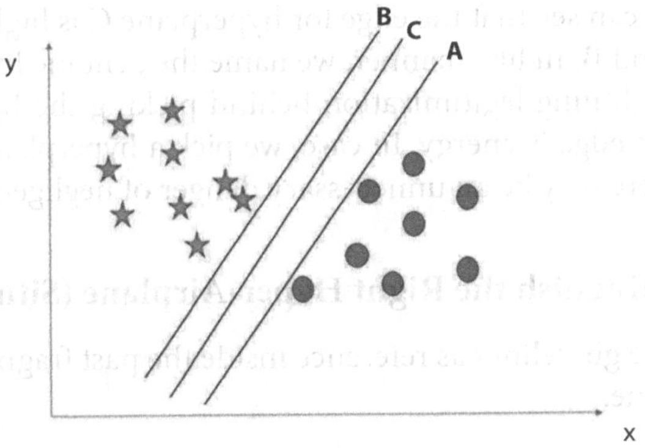

Figure 10.4 Circumstance 2.

Here, enlarging the distances between the closest bits of knowledge point (either heavenliness) and hyperplane will help us with choosing the right hyperplane. This distance is implied as a margin.

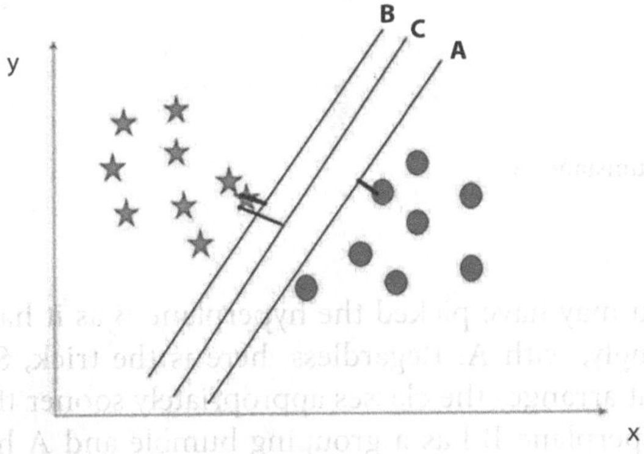

Figure 10.5 Circumstance 2.

Above, you can see that the edge for hyperplane C is high, interestingly, with both A and B. In like manner, we name the genuine hyperplane as C. Some other lightning legitimization behind picking the hyperplane with a predominant edge is energy. In case, we pick a hyperplane having a low edge; then, there may be an unnecessary danger of negligence class.

10.7.3 Distinguish the Right Hyper-Airplane (Situation 3)

Follow: Use the guidelines as reference inside the past fragment to find the right hyperplane.

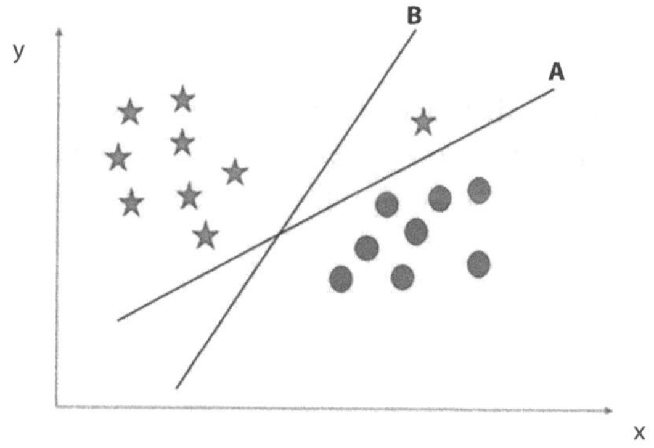

Figure 10.6 Circumstance 3.

Some of you may have picked the hyperplane B as it has an unrivaled edge, interestingly, with A. Regardless, here is the trick, SVM picks the hyperplane that arranges the classes appropriately sooner than expanding edge. Here, hyperplane B has a grouping bumble and A has denoted all precisely. In this way, the right hyperplane is A.

10.7.4 Would We Have the Option to Organize Models (Circumstance 4)?

Under, we are not prepared to disengage the two classes the utilization of a directly line, as one of the stars lies inside the space of the alternative (circle) greatness as a peculiarity.

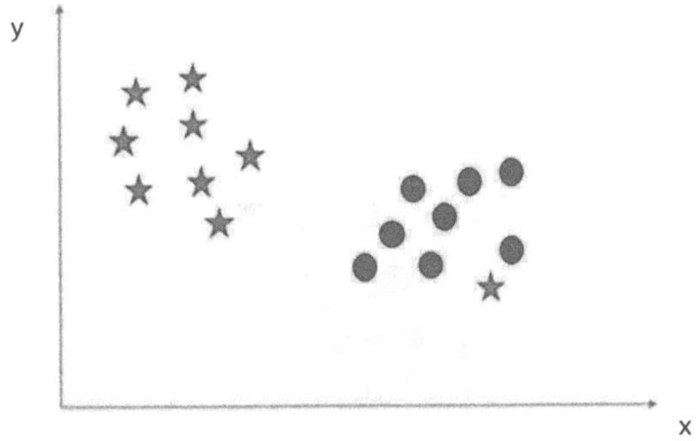

Figure 10.7 Circumstance 4.

As we have, at this point, noted, one significant name at the choice quit takes after a special case for star class. The SVM set of rules has a brand name to neglect oddities and observe the hyperplane that has the most edge. Thus, we can say, SVM class is strong to special cases.

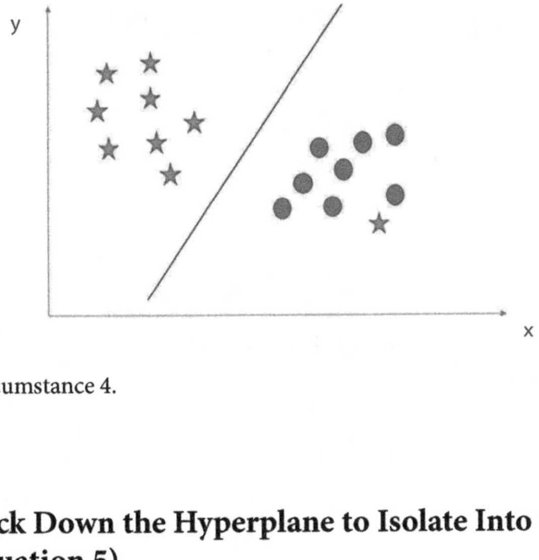

Figure 10.8 Circumstance 4.

10.7.5 Track Down the Hyperplane to Isolate Into Guidelines (Situation 5)

Inside the situation underneath, we cannot have a direct hyper-airplane between the two preparing, as of not long ago, we have least complex checked out the straight hyperplane.

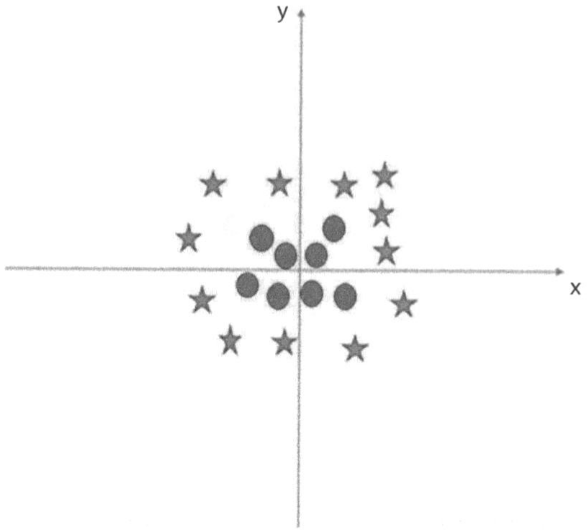

Figure 10.9 Circumstance 5.

SVM can settle this. Without issues! It addresses this difficulty via presenting the extra capacity. Here, we will transfer the shiny new trademark z = x^2 + y^2.

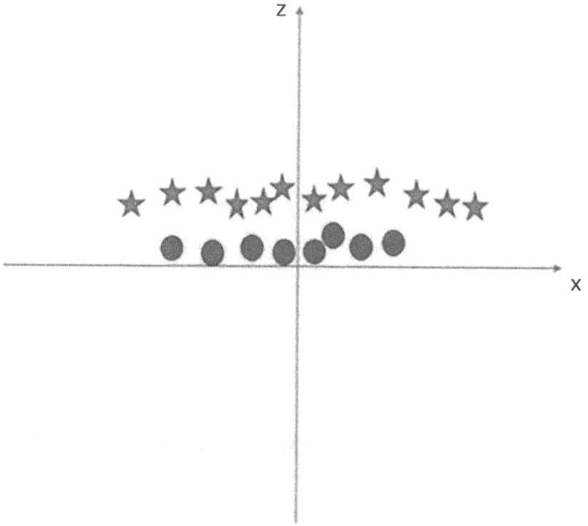

Figure 10.10 Circumstance 5.

Inside the above plot, centers to recall are the following:

- All characteristics for z might be positive typically in light of the truth that z is the squared measure of each x and y.
- In the genuine plot, pink circles appear near the beginning of the x and y hatchets, provoking a decrease cost of z and megastar remarkably far off from the veritable result to the higher worth of z.

Inside the SVM classifier, it is not difficult to have a straight hyperplane between those two getting ready. Regardless, for every single other consuming inquiry that arises, would it be advisable for us that we truly need to add this part genuinely to have a hyperplane? No, the SVM set of rules has a strategy insinuated as the piece stunt. The SVM part is a brand name that takes low layered enter area and changes it to a predominant layeredregion; for example, it changes at this point not distinct issue over to distinguishable issue. It is miles regularly valuable in non-straight division issues. All things considered put, it does a couple of very complicated data changes and, then, at that point, uncovers out the technique to isolate the realities basically based at the names or results you have portrayed.

At the point when we concentrate on the hyperplane in unique entire region maybe a circle.

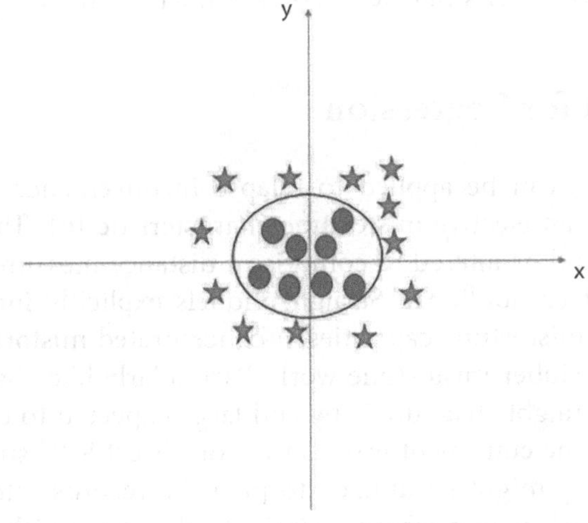

Figure 10.11 Circumstance 5.

Presently, licenses investigate the techniques to apply the SVM classifier set of rules in a data imaginative expertise project [5].

10.8 SVM for Classification

SVM is a valuable methodology for data type. Despite the fact that it is viewed as that neural networks are more straightforward to use than this, however, now and again, inadmissible outcomes are gotten. A class project ordinarily includes preparing and testing real factors that involve two or three records cases. Every model in the tutoring put forth comprises of one objective charge and various characteristics. The objective of SVM is to make a model which predicts the true expense of insights occurrences inside the looking at set that are given best the attributes [6, 7].

Class in SVM is an illustration of supervised getting to know. Realized marks assist with recommending whether or not the machine is acting inside the correct way or not. This insights elements to the favored response, approving the exactness of the framework, or becoming accustomed to help the gadget find ways of acting effectively. A stage in SVM characterization incorporates way of life as that are personally associated with the known examples. That is known as trademark decision or element extraction. Work determination and SVM classification by and large have been utilized in any event, when the expectation of obscure examples is not significant all of the time. They can be utilized to select keysets which are engaged with whatever strategies recognize the training [6, 7].

10.9 SVM for Regression

SVMs, likewise, can be applied to relapse inconveniences through the presentation of an elective misfortune characteristic [6]. The misfortune trademark should be altered to comprise a distance measure. The relapse can be straight or nonlinear. Straight models explicitly incorporate the accompanying misfortune capacities, e-concentrated misfortune abilities, quadratic, and Huber misfortune work. Particularly like characterization issues, a non-straight structure is by and large expected to capably frame records. Inside the equivalent way, as the non-direct SVC strategy, a non-straight planning might be utilized to plan the records into an extreme layered trademark region wherein straight relapse is finished. The piece technique is again utilized to address the scourge of dimensionality. Inside the relapse strategy, there are inconveniences principally dependent on ahead of time comprehension of the issue and the dispersion of the clamor. Inside the shortfall of such insights, Huber's hearty misfortune work has been checked to be an outstanding open door [6, 7].

10.10 Python Implementation of Support Vector Machine

Presently, we can place in power the SVM set of rules the utilization of Python. Here, we will use the vague dataset user_data, which we have utilized in logistic backslide and KNN arrangement [1].

10.10.1 Data Pre-Taking Care of Step

Until the records pre-dealing with step, the code will remain the unclear. Under is the code:

Resulting to executing the underneath code, we will pre-procedure the information. The code will supply the dataset as follows [1].

```
#Data Pre-processing Step
# importing libraries
import numpy as nm
import matplotlib.pyplot as mtp
import pandas as pd

#importing datasets
data_set= pd.read_csv('user_data.csv')

#Extracting Independent and dependent Variable
x= data_set.iloc[:, [2,3]].values
y= data_set.iloc[:, 4].values

# Splitting the dataset into training and test set.
from sklearn.model_selection import train_test_split
x_train, x_test, y_train, y_test= train_test_split(x, y, test_size= 0.25, random_state=0)
#feature Scaling
from sklearn.preprocessing import StandardScaler
st_x= StandardScaler()
x_train= st_x.fit_transform(x_train)
x_test= st_x.transform(x_test)
```

Figure 10.12 Code snippet of data pre-processing.

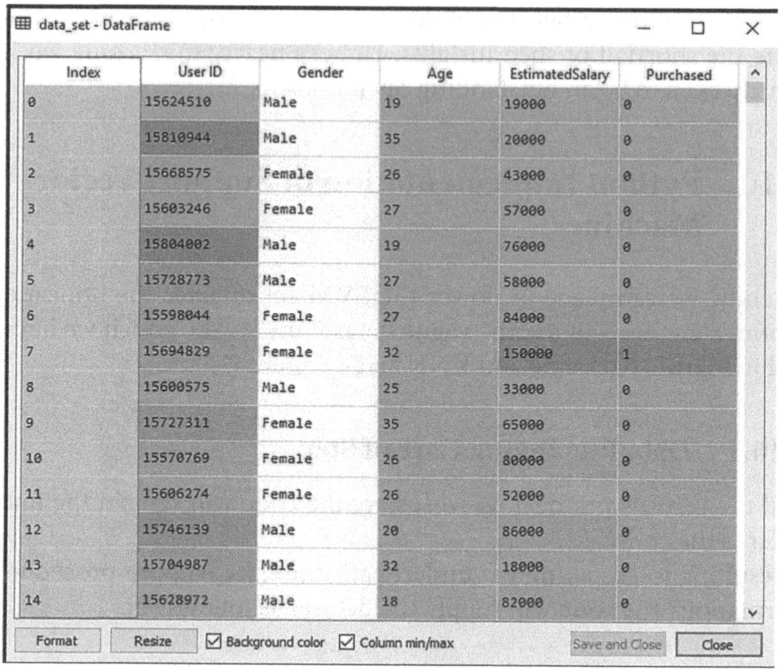

Index	User ID	Gender	Age	EstimatedSalary	Purchased
0	15624510	Male	19	19000	0
1	15810944	Male	35	20000	0
2	15668575	Female	26	43000	0
3	15603246	Female	27	57000	0
4	15804002	Male	19	76000	0
5	15728773	Male	27	58000	0
6	15598044	Female	27	84000	0
7	15694829	Female	32	150000	1
8	15600575	Male	25	33000	0
9	15727311	Female	35	65000	0
10	15570769	Female	26	80000	0
11	15606274	Female	26	52000	0
12	15746139	Male	20	86000	0
13	15704987	Male	32	18000	0
14	15628972	Male	18	82000	0

Figure 10.13 Dataset samples.

The result of the test set would be as follows:

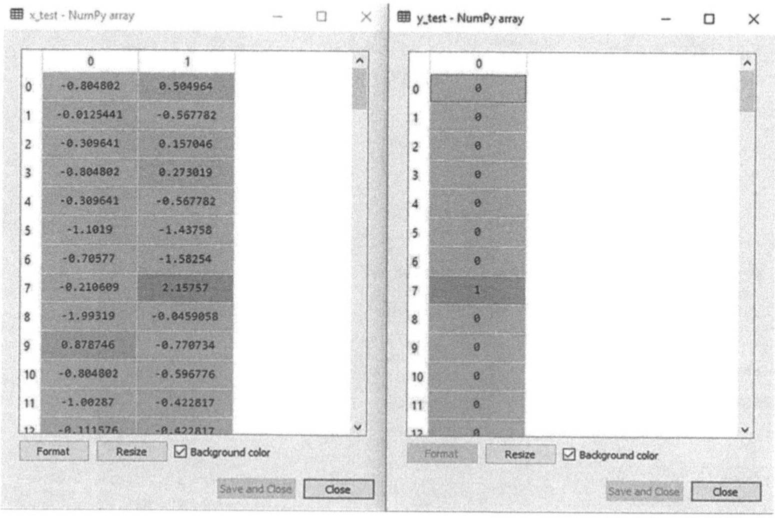

	0	1
0	-0.804802	0.504964
1	-0.0125441	-0.567782
2	-0.309641	0.157046
3	-0.804802	0.273019
4	-0.309641	-0.567782
5	-1.1019	-1.43758
6	-0.70577	-1.58254
7	-0.210609	2.15757
8	-1.99319	-0.0459058
9	0.878746	-0.770734
10	-0.804802	-0.596776
11	-1.00287	-0.422817
12	-0.111576	-0.422817

	0
0	0
1	0
2	0
3	0
4	0
5	0
6	0
7	1
8	0
9	0
10	0
11	0
12	0

Figure 10.14 Test set samples.

10.10.2 Fitting the SVM Classifier to the Readiness Set

By and by, the guidance set may be fit to the SVM. To make the SVM classifier, we can import SVC class from Sklearn.svm library. Under is the best code for it.

```
from sklearn.svm import SVC # "Support vector classifier"
classifier = SVC(kernel='linear', random_state=0)
classifier.fit(x_train, y_train)
```

Figure 10.15 Model fitting (SVM).

In the above code, we have used kernel='linear', as here we are making SVM for straightly separable information. Regardless, we will exchange it for non-straight real factors. Furthermore, we arranged the classifier to the tutoring dataset (x_train, y_train).

10.10.2.1 Outcome

The model presentation might be changed via changing over the worth of C (regularization component), gamma, and part.

```
Out[8]:
SVC(C=1.0, cache_size=200, class_weight=None, coef0=0.0,
    decision_function_shape='ovr', degree=3, gamma='auto_deprecated',
    kernel='linear', max_iter=-1, probability=False, random_state=0,
    shrinking=True, tol=0.001, verbose=False)
```

Figure 10.16 SVM model.

10.10.3 Anticipating the Investigated Set Final Product

Presently, we can anticipate the result for investigate set. For this, we will make another vector y_pred. Under is the best code for it.

```
#Predicting the test set result
y_pred= classifier.predict(x_test)
```

Figure 10.17 Code snippet on prediction.

When you have y_pred vector, then we will think about the outcome of y_pred and y_test to actually take a look at the qualification among the genuine cost and anticipated expense.

10.10.3.1 Yield

Underneath is the result for the forecast of the actually look at set.

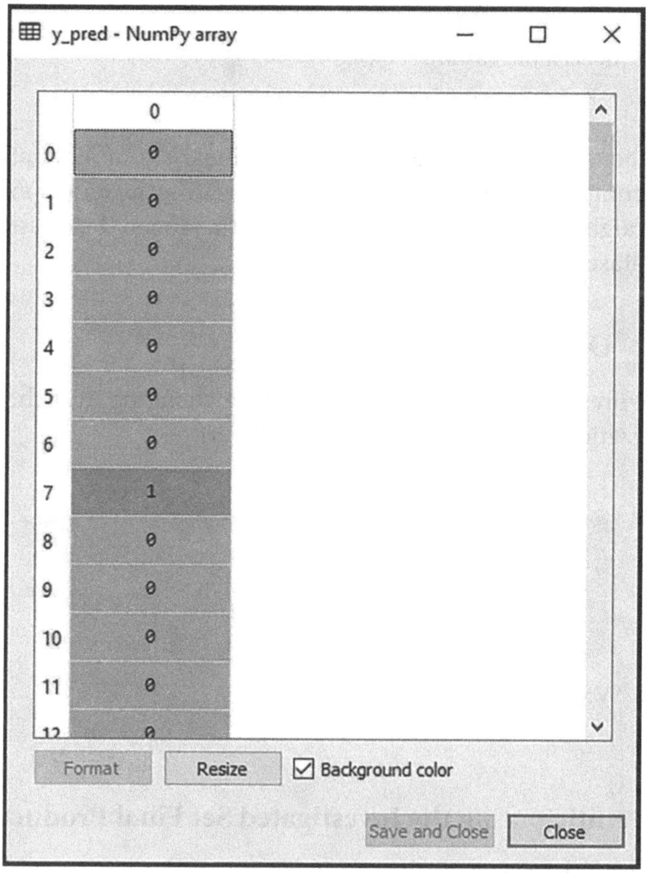

Figure 10.18 Predicted values.

10.10.4 Fostering the Disarray Lattice

Presently, we can see the general exhibition of the SVM that what number of mistaken expectations is there when contrasted with the logistic backslide classifier. To make the chaos framework, we need to import the confusion_matrix part of the sklearn library. Directly following moving the limit, we will name it the usage of another variable cm. The cutoff takes two cutoff points, basically y_true (the genuine attributes) and y_pred (the connected with cost return by utilizing the classifier). Under is an ideal code for it.

```
#Creating the Confusion matrix
from sklearn.metrics import confusion_matrix
cm= confusion_matrix(y_test, y_pred)
```

Figure 10.19 Code snippet for confusion matrix.

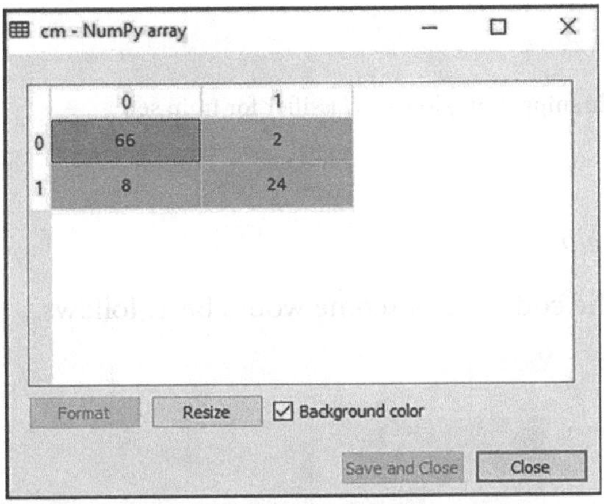

Figure 10.20 Confusion matrix.

As we will see inside the above yield picture, there are 24 + 66 = 90 right guesses and 2 + 8 = 10 right suppositions. Consequently, we can say that our SVM model advanced when appeared differently in relation to the logistic backslide model.

10.10.5 Picturing the Preparation Set Outcome

Presently, we can envision the preparation set outcome, underneath is an ideal code for it.

```
from matplotlib.colors import ListedColormap
x_set, y_set = x_train, y_train
x1, x2 = nm.meshgrid(nm.arange(start = x_set[:, 0].min() - 1, stop = x_set[:, 0].max() + 1, step  =0.01),
nm.arange(start = x_set[:, 1].min() - 1, stop = x_set[:, 1].max() + 1, step = 0.01))
mtp.contourf(x1, x2, classifier.predict(nm.array([x1.ravel(), x2.ravel()]).T).reshape(x1.shape),
alpha = 0.75, cmap = ListedColormap(('red', 'green')))
mtp.xlim(x1.min(), x1.max())
mtp.ylim(x2.min(), x2.max())
for i, j in enumerate(nm.unique(y_set)):
    mtp.scatter(x_set[y_set == j, 0], x_set[y_set == j, 1],
      c = ListedColormap(('red', 'green'))(i), label = j)
mtp.title('SVM classifier (Training set)')
mtp.xlabel('Age')
mtp.ylabel('Estimated Salary')
mtp.legend()
mtp.show()
```

Figure 10.21 Code snippet on plotting classifier for train set.

10.10.5.1 Yield

On executing the code, the outcome would be as follows.

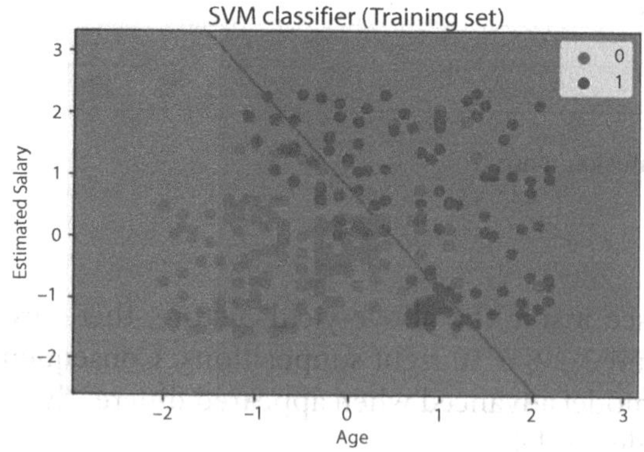

Figure 10.22 Plot on train set SVM classifier.

As it may be obvious, the above yield is appearing comparable as the logistic backslide yield. Inside the outcome, we got the, in a brief moment, line as hyperplane; as a result of the truth, we have included an immediate piece in the classifier. Additionally, we have observed that, besides being inspected over that for the resulting space, the hyperplane in SVM is a quickly line.

10.10.6 Imagining the Investigated Set Outcome

```
#Visulaizing the test set result
from matplotlib.colors import ListedColormap
x_set, y_set = x_test, y_test
x1, x2 = nm.meshgrid(nm.arange(start = x_set[:, 0].min() - 1, stop = x_set[:, 0].max() + 1, step  =0.01),
nm.arange(start = x_set[:, 1].min() - 1, stop = x_set[:, 1].max() + 1, step = 0.01))
mtp.contourf(x1, x2, classifier.predict(nm.array([x1.ravel(), x2.ravel()]).T).reshape(x1.shape),
alpha = 0.75, cmap = ListedColormap(('red','green' )))
mtp.xlim(x1.min(), x1.max())
mtp.ylim(x2.min(), x2.max())
for i, j in enumerate(nm.unique(y_set)):
    mtp.scatter(x_set[y_set == j, 0], x_set[y_set == j, 1],
        c = ListedColormap(('red', 'green'))(i), label = j)
mtp.title('SVM classifier (Test set)')
mtp.xlabel('Age')
mtp.ylabel('Estimated Salary')
mtp.legend()
mtp.show()
```

Figure 10.23 Code snippet on plotting classifier for test set.

10.10.6.1 Yield

On executing the code, the outcome would be as follows.

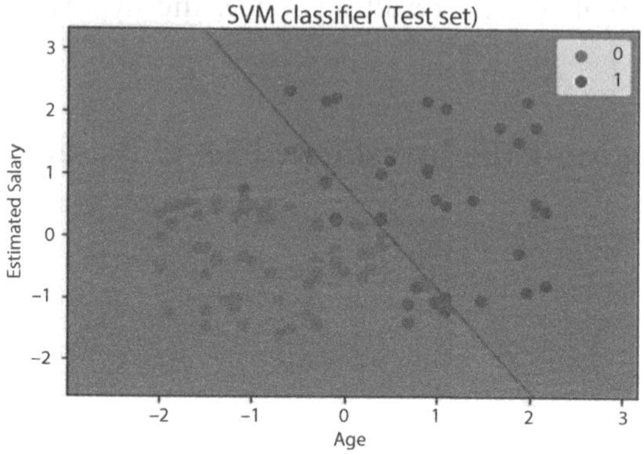

Figure 10.24 Plot on test set SVM classifier.

As we will see inside the above yield photograph, the SVM classifier has confined the customers into two regions (bought or now not purchased). Customers who purchased the SUV are in the red area with the dim red dispersed factors. Furthermore, clients who did not buy the SUV car are inside the unpracticed spot with green dissipated factors. The hyperplane has partitioned the two directions into bought and presently not bought variables [1].

10.10.7 Part or Kernel

We have, as of now, referenced with regard to it. Here, we have different options open with part like, "direct," "rbf," "poly," and others (default cost is "rbf"). Here, "rbf" and "poly" are useful for non-direct hyperplane. License's view the situation, where we have used straight part on component of iris real factor set to sort their class [5].

10.10.8 Support Vector Machine (SVM) Code in Python

Model: Have a straight SVM portion

```
import numpy as np
import matplotlib.pyplot as plt
from sklearn import svm, datasets
```

Figure 10.25 Code snippet for plotting entire SVM classifier with linear kernel.

```
# import some data to play with
iris = datasets.load_iris()
X = iris.data[:, :2] # we only take the first two features. We could
 # avoid this ugly slicing by using a two-dim dataset
y = iris.target
```

Figure 10.26 (Continuation) Code snippet for plotting entire SVM classifier with linear kernel.

```
# we create an instance of SVM and fit out data. We do not scale our
# data since we want to plot the support vectors
C = 1.0 # SVM regularization parameter
svc = svm.SVC(kernel='linear', C=1,gamma=0).fit(X, y)
```

Figure 10.27 (Continuation) Code snippet for plotting entire SVM classifier with linear kernel.

```
# create a mesh to plot in
x_min, x_max = X[:, 0].min() - 1, X[:, 0].max() + 1
y_min, y_max = X[:, 1].min() - 1, X[:, 1].max() + 1
h = (x_max / x_min)/100
xx, yy = np.meshgrid(np.arange(x_min, x_max, h),
 np.arange(y_min, y_max, h))
```

Figure 10.28 (Continuation) Code snippet for plotting entire SVM classifier with linear kernel.

```
plt.subplot(1, 1, 1)
Z = svc.predict(np.c_[xx.ravel(), yy.ravel()])
Z = Z.reshape(xx.shape)
plt.contourf(xx, yy, Z, cmap=plt.cm.Paired, alpha=0.8)
```

Figure 10.29 (Continuation) Code snippet for plotting entire SVM classifier with linear kernel.

```
plt.scatter(X[:, 0], X[:, 1], c=y, cmap=plt.cm.Paired)
plt.xlabel('Sepal length')
plt.ylabel('Sepal width')
plt.xlim(xx.min(), xx.max())
plt.title('SVC with linear kernel')
plt.show()
```

Figure 10.30 (Continuation) Code snippet for plotting entire SVM classifier with linear kernel.

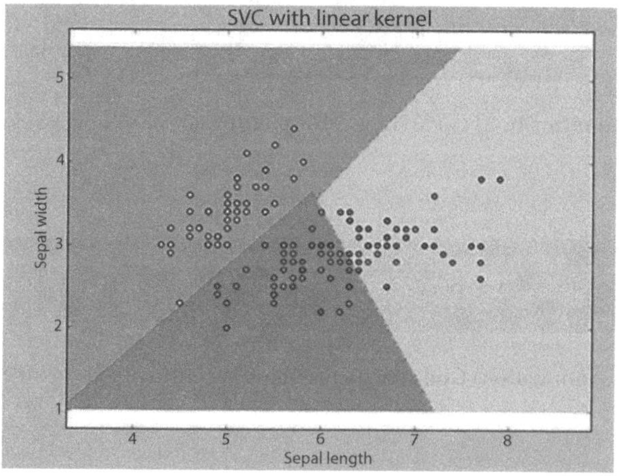

Figure 10.31 Plot on test set SVM with linear kernel.

Model: Use SVM rbf part

Substitute the part type to rbf in under line and notice the effect.

```
svc = svm.SVC(kernel='rbf', C=1,gamma=0).fit(X, y)
```

Figure 10.32 Code snippet for using SVM rbf part.

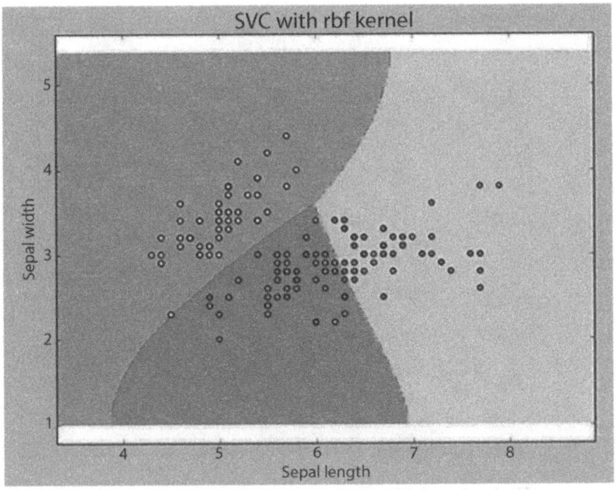

Figure 10.33 Plot on SVM with rbf kernel.

We recommend you to pass for direct SVM bit assuming that you have an enormous scope of capacities (greater than a thousand) because of the reality that, almost certainly, the insights is straightly detachable in extreme layered space. Additionally, you could utilize RBF anyway recall to cross-support for its limits to abstain from over-becoming.

Gamma: Piece coefficient for "rbf," "poly," and "sigmoid." Higher the expense of gamma, will attempt to address sound the as consistent with preparing informational collection for example speculation errors and rationale over-transforming into hassle [5].

Model: We should contrast assuming that we have gamma explicit gamma esteems like zero, 10 or 100.

```
svc = svm.SVC(kernel='rbf', C=1,gamma=0).fit(X, y)
```

Figure 10.34 Code snippet on SVM with rbf kernel for gamma = 0.

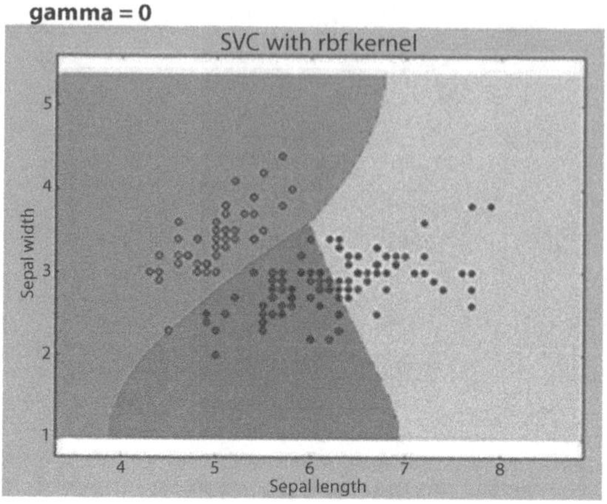

Figure 10.35 Plot on SVM with rbf kernel for gamma = 0.

```
svc = svm.SVC(kernel='rbf', C=100,gamma=10).fit(X, y)
```

Figure 10.36 Code snippet on SVM with rbf kernel for gamma = 10 and c = 100.

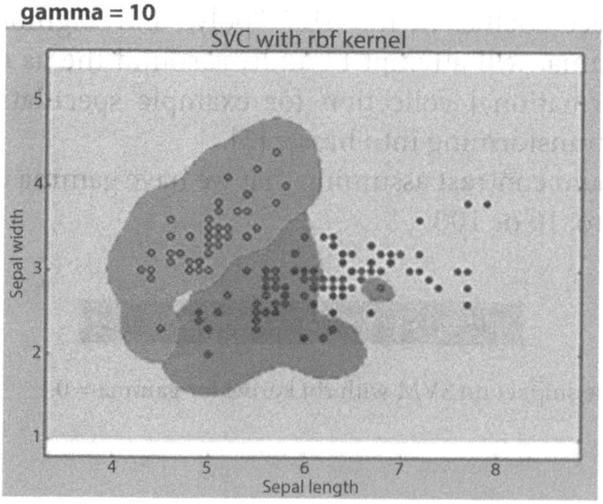

Figure 10.37 Plot on SVM with rbf kernel for gamma = 10.

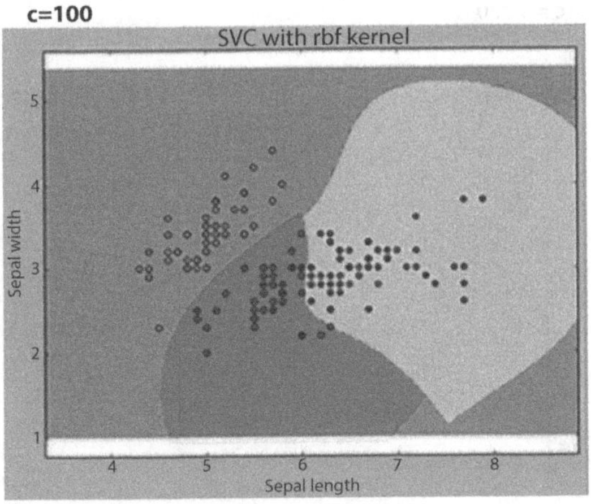

Figure 10.38 Plot on SVM with rbf kernel for c = 100.

```
svc = svm.SVC(kernel='rbf', C=1000,gamma=100).fit(X, y)
```

Figure 10.39 Code snippet on SVM with rbf kernel for gamma = 100 and c = 1,000.

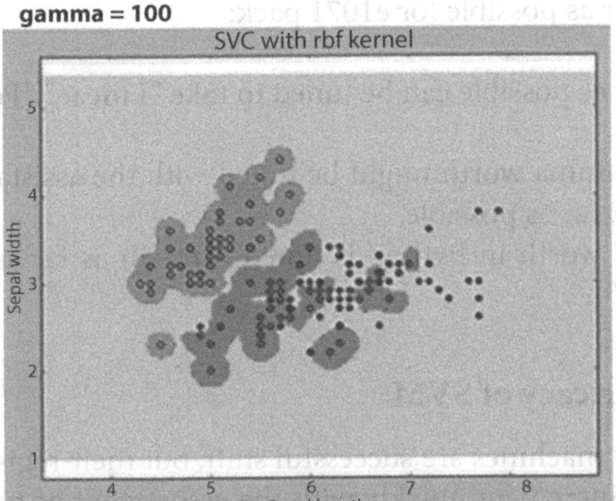

Figure 10.40 Plot on SVM with rbf kernel for gamma = 100.

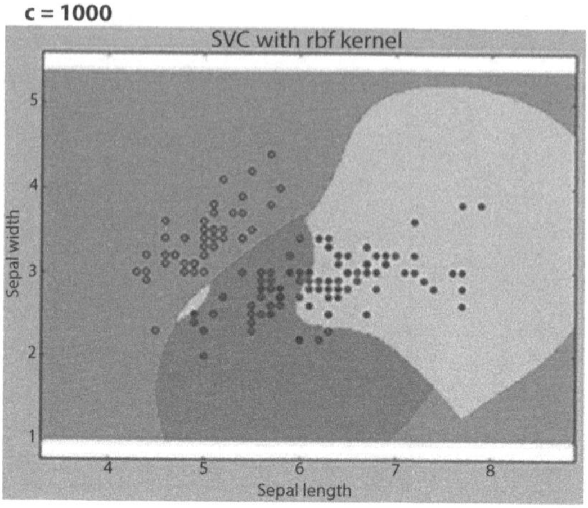

Figure 10.41 Plot on SVM with rbf kernel for c = 1,000.

C: Penalty limit C of the botch term. It likewise controls the substitute between clean decision limitations and gathering the planning factors feasibly.

We want to ceaselessly look at the pass-endorsement measure to have convincing blend of those limits and abstain from over-becoming.

In R, SVMs could be tuned moreover as they're in Python. Recommended under are as far as possible for e1071 pack:

- As far as possible can be tuned to take "Linear," "Poly," "rbf," etc.
- The gamma worth might be tuned with the assistant of setting as far as possible.
- The C worth in Python is tuned through as far as possible in R [5].

10.10.9 Intricacy of SVM

Support vector machines are successful stuff, but their registry and capacity necessities increment quickly with the assortment of tutoring vectors. The center of a SVM is a quadratic programming issue (QP), isolating aide vectors from the unwinding of the schooling information. The QP solver utilized by the libsvm-based execution scales among O(m_features * m_samples^2) and O(m_features * m_samples^3) contingent upon how strongly the libsvm store is utilized by and by (dataset based). Assuming the information could be exceptionally meager need to get supplanted by the normal scope of non-0 abilities in an example vector.

For the straight case, the arrangement of rules used in LinearSVC by the liblinear execution is considerably greener than its libsvm-based SVC partner and may scale directly to a large number of tests as well as functions [2].

References

1. https://www.javatpoint.com/machine-learning-support-vector-machine-algorithm
2. https://scikit-learn.org/stable/modules/svm.html
3. https://en.wikipedia.org/wiki/Support-vector_machine
4. https://www.geeksforgeeks.org/support-vector-machine-algorithm/
5. https://www.analyticsvidhya.com/blog/2017/09/understaing-support-vector-machine-example-code/
6. https://www.researchgate.net/publication/37419815_An_introduction_to_support_vector_machines_and_other_kernel-based_learning_methods_Repr
7. https://course.ccs.neu.edu/cs5100f11/resources/jakkula.pdf

11
Clustering

Clustering is the task of separating the populace or realities factors into a few organizations to such an extent that information factors inside similar organizations are more like diverse data factors inside a similar gathering and unmistakable to the records focuses in different companies. It is an assortment of devices dependent on comparability and divergence among them.

It is miles a kind of solo getting to know approach. A solo acquiring information on procedure is a manner by which we draw references from datasets on the whole with input data without arranged reactions. For the most part, it is far utilized as a technique to find critical construction, logical basic methodologies, generative highlights, and groupings intrinsic in a bunch of models [1].

11.1 Example

Permit perceives the grouping strategy with this present reality occasion of mall: While we go to any retail plaza, we will concentrate on that the issues with practically identical use are assembled. Alongside, the shirts are assembled in one stage, and the pants are in various areas; furthermore, in vegetable segments, apples, bananas, mangoes, and so on, are gathered in independent segments, all together that we are fit for without trouble find the issues. The bunching approach additionally works similarly. Various instances of bunching are gathering records in accordance with the subject.

The grouping strategy can be generally utilized in different obligations. A few greatest ordinary employments of this strategy are as follows:

- Market segmentation
- Measurable data investigation
- Social people group examination

Kolla Bhanu Prakash. Machine Learning for Industrial Applications, (177–204) © 2024 Scrivener Publishing LLC

- ➢ Photo division
- ➢ Irregularity location, thus far.

Beside those general uses, it is miles used by Amazon in its idea device to give the pointers as shown by the past search for of item. Netflix additionally utilizes this way to deal with encourage movies and net series to its clients as in sync with the watch history.

The underneath graph clarifies the running of the grouping calculation. We can see that the outstanding perfection is separated into various associations with equivalent houses [2].

11.2 Kinds of Clustering

Widely speaking, bunching might be isolated into two subgroups as follows.

11.2.1 Hard Clustering

In intense bunching, each data point both has a place with a group totally or presently not. For instance, in the above model, each customer is set into one gathering out of the 10 associations.

11.2.2 Delicate Clustering

In delicate grouping, instead of setting every data figure right in a different bunch, a chance of that record component to be inside the one's bunches is

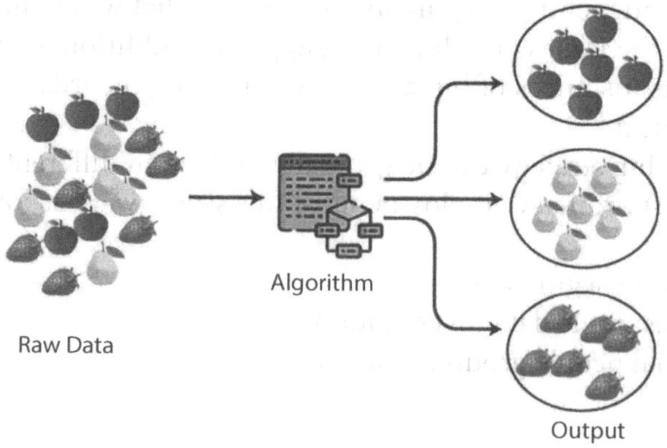

Figure 11.1 Clustering on fruits.

allocated. For instance, from the above situation, every buyer is allocated a chance to be in all of the 10 bunches of the retail saved [3].

Regardless, there are, moreover, other grouped procedures of clustering that exist. The following are the essential grouping strategies used in gadget examining:

1. Partitioning clustering
2. Density-based thoroughly clustering
3. Distribution model basically based clustering
4. Hierarchical clustering
5. Fuzzy clustering

11.2.2.1 Dividing Clustering or Partitioning Clustering

It is a sort of grouping that partitions current realities into non-various leveled partnerships. It is additionally called the centroid-fundamentally based methodology. The most commonplace instance of dividing is the k-suggest clustering estimation.

In this sort, the dataset is parted into a fixed of alright organizations, in which alright is utilized to characterize the scope of pre-characterized organizations. The bunch community is made in this kind of way that the space between the measurements points of 1 group is negligible in contrast with each and every other group centroid.

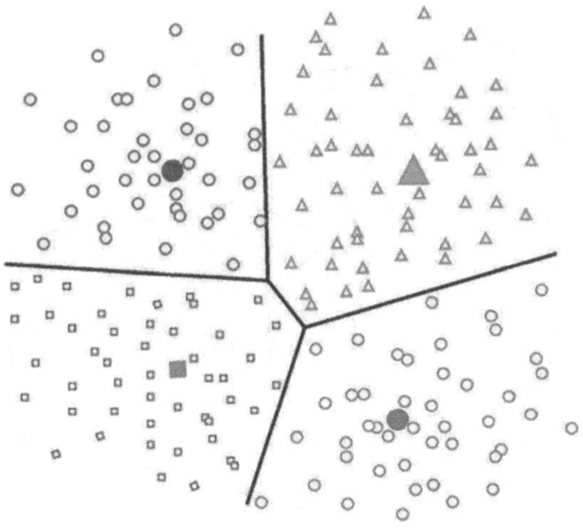

Figure 11.2 Partitioning clustering.

11.2.2.2 Thickness Essentially Based Clustering or Density Fundamentally Based Clustering

The thickness essentially based gathering procedure interfaces the particularly thick locale into bunches, and the self-assertively molded dispersions are shaped inasmuch as the thick area might be associated. This arrangement of rules does it through sorting out unprecedented bunches in the dataset and interfacing the areas of extreme densities into groups. The thick regions in realities space are separated from each other with the guide of way of sparser locales.

These calculations can confront inconvenience in grouping the record that it focuses if the dataset has different densities and inordinate aspects.

11.2.2.3 Transport Model–Based Clustering or Distribution Model–Based Clustering

Inside the transport transformation fundamentally based gathering methodology, the information is isolated basically subject to the shot at the way a dataset has a spot with a picked flow. The gathering is performed by utilizing and accepting a few circulations like the normal Gaussian distribution.

An occasion of this sort is the anticipation maximization clustering set of decides that utilizes Gaussian total designs (GMM).

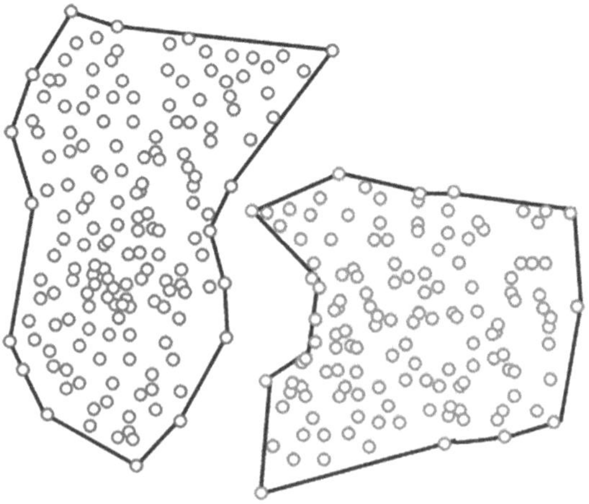

Figure 11.3 Density fundamentally based clustering.

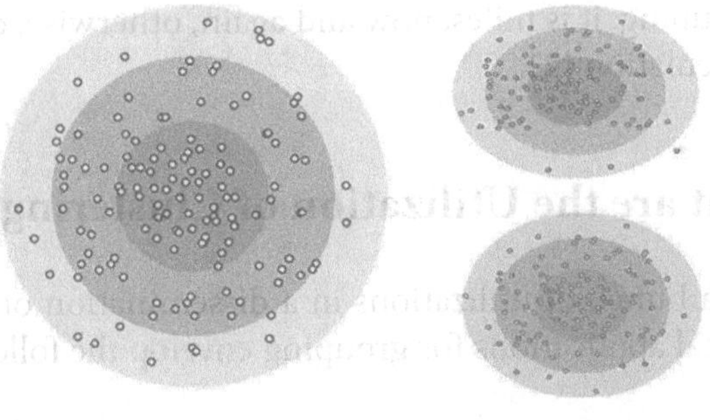

Figure 11.4 Distribution model-based clustering.

11.2.2.4 Progressive Clustering or Hierarchical Clustering

Progressive grouping can be utilized as a chance for separated grouping as there is no essential of pre-showing how much packs to be made. On this technique, the dataset is partitioned into bunches to make a tree-like shape, which is, moreover, known as a dendrogram. The perceptions or any amount of groups can be settled on via cutting the tree at the suitable degree. The most normal illustration of this strategy is the agglomerative hierarchical arrangement of rules.

11.2.2.5 Fluffy Clustering or Fuzzy Clustering

Fluffy bunching is a kind of smooth strategy wherein an insight item may, furthermore, have a place with various gatherings or groups. Each dataset has a bunch of participation coefficients, which rely on the certificate of club to be in a group. The ragged C-way set of rules is an illustration of

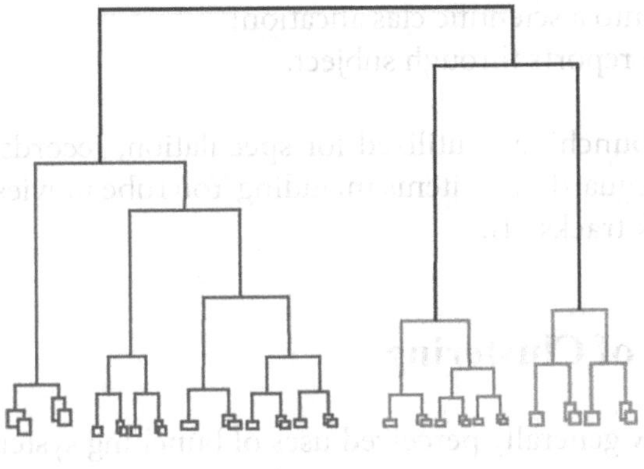

Figure 11.5 Distribution model-based clustering.

this sort of grouping; it is miles, now and again, otherwise, called the thick alright way calculation [2].

11.3 What are the Utilization of Clustering?

Grouping has a bunch of utilizations in a dissemination of enterprises. A couple of normal applications for grouping envelop the following:

- ✓ commercial center division
- ✓ social local area assessment
- ✓ search outcome gathering
- ✓ clinical imaging
- ✓ photograph division
- ✓ oddity identification

Subsequent to grouping, each bunch is relegated a number called a bunch personality. Presently, you could consolidate the whole trademark set for an occasion into its group ID. Addressing a complicated case through a basic group personality makes bunching strong. Broadening the idea, bunching records can improve monstrous datasets.

As an occasion, you can associate objects via phenomenal capacities as exhibited in the accompanying models.

11.4 Models

- ➢ foundation stars by brilliance;
- ➢ bunch creatures through hereditary measurements legitimate into a scientific classification;
- ➢ bunch reports through subject.

In Google, bunching is utilized for speculation, records pressure, and privateness safeguarding in items including YouTube movies, Play applications, and track tracks [4].

11.5 Uses of Clustering

Under are a few generally perceived uses of bunching systems in contraption learning.

11.5.1 In Character of Most Tumor Cells

The grouping calculations are comprehensively utilized for the personality of carcinogenic cells. It splits the destructive and non-carcinogenic data units into one between a thoughtful organizations.

11.5.2 In Web Crawlers Like Google

Web crawlers like Google and Yippee are also works of art at the grouping approach. The output shows up essentially based at the nearest thing to the journey question. It does it via gathering comparable record contraptions in a single gathering; this is a long way from the other numerous devices. The right consequence of an inquiry depends at the excellent of the grouping set of rules utilized.

11.5.3 Shopper Segmentation

It is miles utilized in commercial center examination to stage the clients essentially dependent on their inclination and inclinations.

11.5.4 In Biology

It is utilized inside the science stream to group unique types of blossoms and creatures utilizing the image notoriety method.

11.5.5 In Land Use

The bunching strategy is used in distinguishing the district of tantamount grounds utilized in the GIS information base. This might be extremely helpful to find why the particular land should be utilized, which means, for which reason, it is miles more appropriate [2].

11.6 Bunching Algorithms or Clustering Algorithms

The clustering calculations can be partitioned basically based at the styles that are characterized previously. There are select sorts of bunching calculations posted, but only a couple are generally utilized. The grouping set of rules is basically founded on the kind of information that we are the utilization of. For example, a few calculations need to risk everything and the kitchen sink of groups inside the given dataset, simultaneously as a couple are needed to find the base distance among the perception of the dataset.

Here, we are examining fundamentally well known clustering calculations that are extensively utilized in AI.

11.6.1 K-Means Clustering

K-means may be the most notable bunching set of arrangements. It is instructed in a ton of early-on data innovative ability and gadget acquiring information on training. It is ideal to appreciate and put in power in code! Explore the photo under for a portrayal.

K-implies has an increase that it is very quick, as everything we are doing is processing the distances among variables and gathering focuses a couple of calculations! It hence has a direct intricacy O(n).

Then again, k-way has two or three dangers. For one thing, you want to choose the number of associations/examples there are. This is not inconsequential, and, ideally, with a grouping calculation, we would require it to parent the ones out for us on the grounds that the variable of it is miles to acquire a few discernments from the records. K-implies additionally begins with an irregular longing of bunch offices, and, hence, it can yield unmistakable grouping consequences for explicit runs of the arrangement of rules. Thus, the outcomes probably would not be repeatable and nonattendance consistency.

11.6.2 Mean-Shift Clustering

Mean shift grouping is a sliding window-based absolutely calculation that endeavors to find thick spaces of records factors. It is far a centroid-based arrangement of rules that implies that the objective is to track down the center components of every foundation class, which matches by refreshing candidates for focus elements to be the propose of the focuses inside the sliding window. Those up-and-comer windows are then separated in a distribute handling degree to remove near copies, shaping the last arrangement of focus factors and their comparing organizations.

As opposed to k-mean grouping, there might be no need to pick the scope of bunches as infer shift precisely finds this. The way that the bunch offices merge nearer to the elements of most thickness is in like manner pretty legitimate as it is far really natural to perceive and fits appropriately in a positively data-driven sense. The disadvantage is that the decision of the window length/sweep "r" can be non-insignificant.

11.6.3 Thickness or Density-Based Spatial Clustering of Application with Noise (DBSCAN)

DBSCAN is a thickness-based completely grouped calculation like mean-shift, but with numerous amazing benefits. Look at another extravagant realistic underneath and how about we get started!

DBSCAN represents a couple of top notch benefits over other grouping calculations. First and foremost, in all actuality, it does do not really need a pe-set wide assortment of bunches by any means. It also distinguishes anomalies as commotions, as opposed to mean-shift that, as a general rule, tosses them directly into a bunch in any event, assuming the record element might be altogether different. Besides, it can find discretionarily estimated and subjectively normal bunches pretty appropriately.

The fundamental disadvantage of DBSCAN is that it does not complete further to others even as the groups are of different thickness. This is because of the truth that putting the distance limit ε and pinpointing for the assortment of the local area components will fluctuate from one group to another while the thickness changes. This disadvantage furthermore occurs with extremely high-layered records thinking about that again the hole limit ε will become intense to appraise.

11.6.4 Assumption Maximization Clustering Utilizing Gaussian Combination Models

One of the significant disadvantages of k-approach is its credulous utilization of the mean incentive for the group community. We can see the reason why this is not the fantastic method of getting things done via looking at the image beneath. At the left-hand viewpoint, it appears to be very obvious to the natural eye that there are round bunches with explicit span's designated on the equivalent infer. K-mean cannot deal with this because of the reality that the recommended upsides of the groups are extremely near one another. K-mean, moreover, comes up short in cases in which the bunches are not roundabout, again as a result of the utilization of the recommend as group center.

Gaussian total models (GMMs) supply us more noteworthy adaptability than k-mean. With GMMs, we expect that the reality focuses are Gaussian-dispensed; this is a mile less prohibitive presumption than articulating that they might be roundabout by utilizing the propose. That way, we have boundaries to clarify the state of the bunches: the propose and the normal, worn out deviation! Taking a model in aspects, on account of this, the bunches can take any sort of curved structure. Hence, each Gaussian appropriation is allocated to an unmarried group.

There are two critical benefits to the utilization of GMMs. Leading GMMs are stacks more adaptable in expressions of group covariance than k-implies; because of the standard deviation boundary, the bunches can handle any oval structure, rather than being restricted to circles. K-strategy is an exceptional instance of GMM wherein each group's covariance along all aspects of the technique is zero. Furthermore, when you consider that GMMs use probabilities, they can have several groups for every data point. So assuming that an insights factor is inside the center of two covering bunches, we can, as a general rule, characterize its radiance through reporting that it has a place X-rate to style 1 and Y-percent to greatness 2, i.e., GMMs guide blended club.

11.6.5 Agglomerative Hierarchical Clustering

Progressive bunching calculations fall into two classes: apex down or rear up. Rear-up calculations treat each record factor as an unmarried bunch on the start after which progressively union (or agglomerate) sets of groups till all bunches had been converged into a solitary group that conveys all data focuses. Base-up various leveled grouping is, hence, alluded to as progressive agglomerative bunching or HAC. This pecking order of groups is addressed as a tree (or dendrogram). The establishment of the tree is the particular bunch that assembles every one of the examples, the leaves being the groups with least difficult one example. Try out the picture underneath for an occurrence prior to moving immediately to the calculation steps

Different evened out batching does not anticipate that we should decide the kind of gatherings, and we will even pick which colossal type of groups appears to be extraordinary on the grounds that we are developing a tree. Also, the arrangement of rules is not touchy to the decision of distance metric; every one of them tends to canvas what is more pleasantly though; with remarkable grouping calculations, the decision of distance metric is critical. An especially evident use instance of progressive grouping techniques is that, while the basic measurements have a various leveled shapes and, furthermore, you need to improve the pecking order, astonishing bunching calculations cannot do this. Those gifts of various leveled groupings come at the charge of abatement generally speaking execution, because it has a period intricacy of $O(n^3)$, dislike the straight intricacy of alright propose and GMM [5].

11.7 Instances of Clustering Algorithms

1. Library installation
2. Clustering dataset
3. Affinity propagation
4. Agglomerative clustering
5. BIRCH
6. DBSCAN
7. K-means
8. Mini-batch k-means
9. Mean shift
10. OPTICS
11. Spectral clustering
12. Gaussian mixture model [6]

11.7.1 Library Setup

To begin with, permit introduces the library.

Try not to pass this progression as you might have to guarantee you have the current-day model setup.

You could introduce the scikit-inspect library utilizing the pip Python installer, as follows:

```
sudo pip install scikit-learn
```

Then, permit checks that the library is setup and you are utilizing a state-of-the-art form.

Run the resulting content to print the library model reach.

```
#check scikit-learn version
import sklearn
print(sklearn.__version__)
```

Running the model, you should see the ensuing rendition range or higher.

```
0.22.1
```

Figure 11.6 Scikit-learn version.

11.7.2 Grouping or Clustering Dataset

We can utilize the make_classification() capacity to make an investigate parallel class dataset.

The dataset might have 1,000 models, with input abilities and one group predictable with class. The bunches are outwardly evident in aspects all together that we can plot current realities with a disperse plot and shade the variables inside the plot through the appointed group. This might assist with seeing, basically at the investigate bother, how "pleasantly" the bunches were analyzed.

An occasion of making and summing up the engineered bunching dataset is recorded beneath.

```
#synthetic classification dataset
from numpy import where
from sklearn.datasets import make classification
from matplotlib import pyplot
# define dataset
x, y = make_classification(n_samples=1000, n_features=2, n_informative-2, n_redundant-0, n_clusters_per_class=1, random_state=4)
# create scatter plot for samples from each class
for class value in range(2):
	# get row indexes for samples with this class
	row_ix = where(y = class_value)
	# create scatter of these samples
	pyplot.scatter (X[row_ix, 0], x[row_ix, 1])
# show the plot
pyplot.show()
```

11.7.3 Fondness or Affinity Propagation

Affection propagation fuses seeing as a fixed of models that fine summarize current real factors.

The whole model is listed underneath.

```
#affinity propagation clustering
from numpy import unique
from numpy import where
from sklearn.datasets import make_classification
```

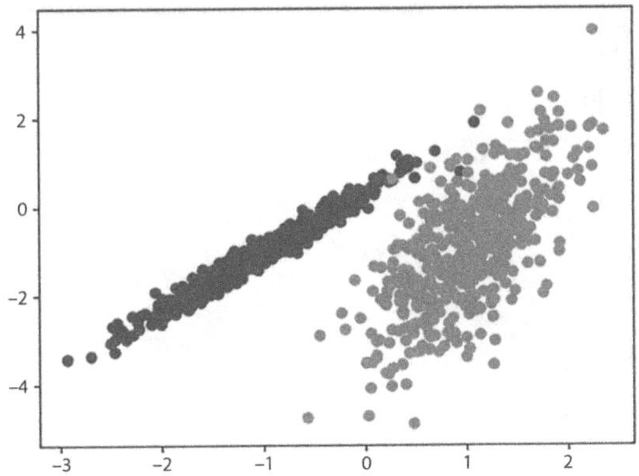

Figure 11.7 Scatter plot on clustering.

```
from sklearn.cluster import AffinityPropagation
from matplotlib import pyplot
# define dataset
X, _ = make_classification(n_samples-1000, n_features=2, n_informa-
tive-2, n_redundant-0, n_clusters_per_class-1, random_state=4)
# define the model
Model=AffinityPropagation(damping-0.9)
# fit the model
model.fit(X)
# assign a cluster to each example
yhat=model.predict(X)
# retrieve unique clusters
clusters unique(yhat)
# create scatter plot for samples from each cluster
for cluster in clusters:
        # get row indexes for samples with this cluster
        row_ix = where(yhat == cluster)
        # create scatter of these samples
        pyplot.scatter(x[row_ix, 0], x[row_ix, 1])
#show the plot
pyplot.show()
```

11.7.4 Agglomerative Clustering

Agglomerative bunching involves consolidating models till the favored assortment of groups is completed.

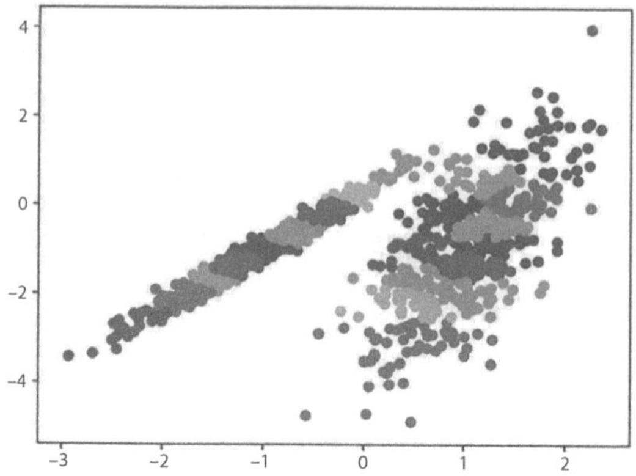

Figure 11.8 Scatter plot on affinity propagation.

The total model is recorded under.

```
#agglomerative clustering
from numpy import unique
from numpy import where
from sklearn.datasets import make_classification
from sklearn.cluster import AgglomerativeClustering
from matplotlib import pyplot
# define dataset
X, = make_classification(n_samples-1000, n_features=2, n_informative-2,
n_redundant-0, n_clusters_per_class=1, random_state=4)
# define the model
model = AgglomerativeClustering(n_clusters=2)
# fit model and predict clusters.
yhat = model.fit_predict(X)
# retrieve unique clusters
clusters unique(yhat)
# create scatter plot for samples from each cluster
for cluster in clusters:
# get row indexes for samples with this cluster
row_ix=where(yhat == cluster)
#create scatter of these samples
pyplot.scatter (x[row_ix, 0], x[row_ix, 1])
# show the plot
pyplot.show()
```

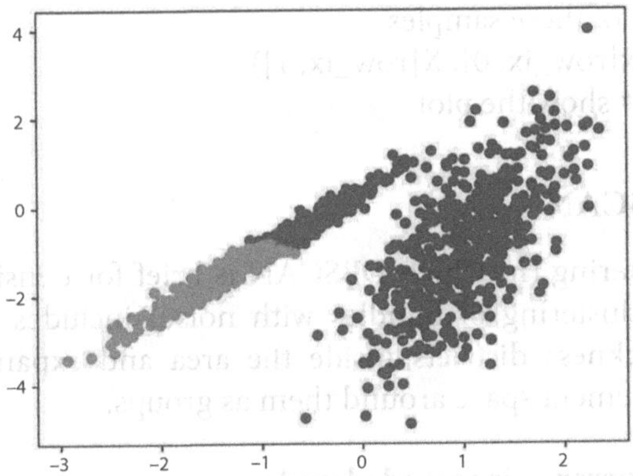

Figure 11.9 Scatter plot on agglomerative clustering.

11.7.5 BIRCH

BIRCH clustering (BIRCH is short for balanced iterative bringing down and clustering the utilization of hierarchies) includes developing a tree shape from which group centroids are extricated.

The entire model is recorded underneath.

```
# birch clustering
from numpy import unique
from numpy import where
from sklearn.datasets import make_classification
from sklearn.cluster import Birch
from matplotlib import pyplot
# define dataset
X, _ = make_classification(n_samples-1000, n_features=2, n_informa-
tive-2, n_redundant-0, n_clusters_per_class-1, random_state=4)
# define the model
model= Birch(threshold-0.01, n_clusters=2)
# fit the model
model.fit(X)
yhat = model.predict(X)
# retrieve unique clusters
clusters = unique(yhat)
# create scatter plot for samples from each cluster
for cluster in clusters:
row_ix = where(yhat cluster) #get row indexes for samples with this cluster
```

```
# create scatter of these samples
pyplot.scatter(x[row_ix, 0], X[row_ix, 1])
pyplot.show() # show the plot
```

11.7.6 DBSCAN

DBSCAN clustering (in which DBSCAN is brief for density-based absolutely spatial clustering of bundles with noise) includes tracking down inordinate thickness districts inside the area and expanding the one's spaces of the element space around them as groups.

The entire occurrence is recorded under.

```
# dbscan clustering
from numpy import unique
from numpy import where
from sklearn.datasets import make_classification
from sklearn.cluster import DBSCAN
from matplotlib import pyplot
# define dataset
X, = make_classification(n_samples-1000, n_features=2, n_informative-2,
n_redundant-0, n_clusters_per_class-1, random_state=4)
# define the model
model = DBSCAN(eps=0.30, min_samples=9)
# fit model and predict clusters
yhat = model.fit_predict(X)
# retrieve unique clusters
```

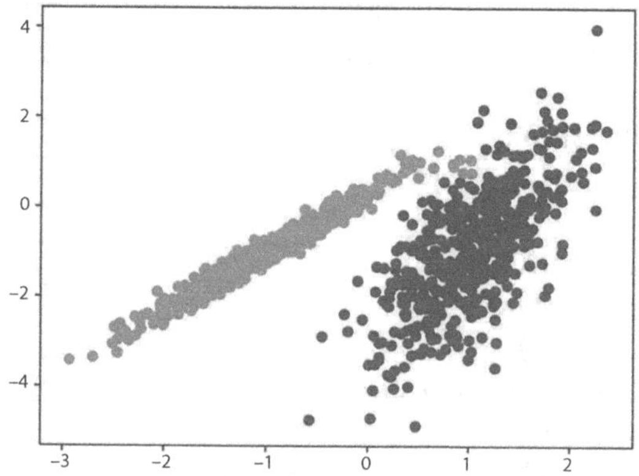

Figure 11.10 Scatter plot on BIRCH clustering.

```
clusters =unique(yhat)
# create scatter plot for samples from each cluster
for cluster in clusters:
# get row indexes for samples with this cluster
row_ix = where(yhat cluster)
# create scatter of these samples
pyplot.scatter(x[row_ix, 0], x[row_ix, 1])
# show the plot
pyplot.show()
```

11.7.7 K-Means

K-means clustering might be the greatest generally perceived bunching set of rules and incorporates appointing guides to groups that permits you to diminish the change inside each bunch.

The total case is recorded underneath.

```
# k-means clustering
from numpy import unique
from numpy import where
from sklearn.datasets import make_classification
from sklearn.cluster import KMeans
from matplotlib import pyplot
X, = make_classification(n_samples=1000, n_features=2, n_informative-2,
n_redundant-0, n_clusters_per_class=1, random_state=4) # define dataset
model= KMeans (n_clusters=2)
```

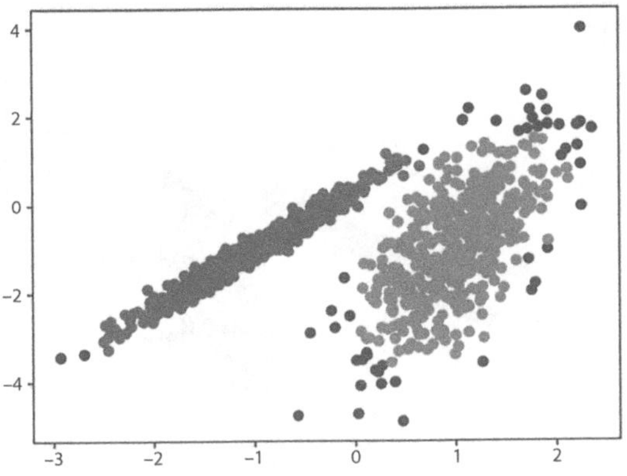

Figure 11.11 Scatter plot on DBSCAN.

```
# fit the model
model.fit(X)
# assign a cluster to each example
yhat= model.predict(X)
# retrieve unique clusters
clusters unique(yhat)
# create scatter plot for samples from each cluster
for cluster in clusters:
# get row indexes for samples with this cluster
row_ix where(yhat == cluster)
# create scatter of these samples
pyplot.scatter(x[row_ix, 0], x[row_ix, 1])
pyplot.show() # show the plot
```

11.7.8 Mini-Batch K-Means

Mini-batch k-means is an adjusted model of k-way that makes updates to the bunch centroids that use scaled down clusters of tests rather than the entire dataset, which could make it quicker for colossal datasets and perhaps more noteworthy solid to factual commotion.

The entire occurrence is ordered under.

```
# mini-batch k-means clustering
from numpy import unique
from numpy import where
from sklearn.datasets import make_classification
```

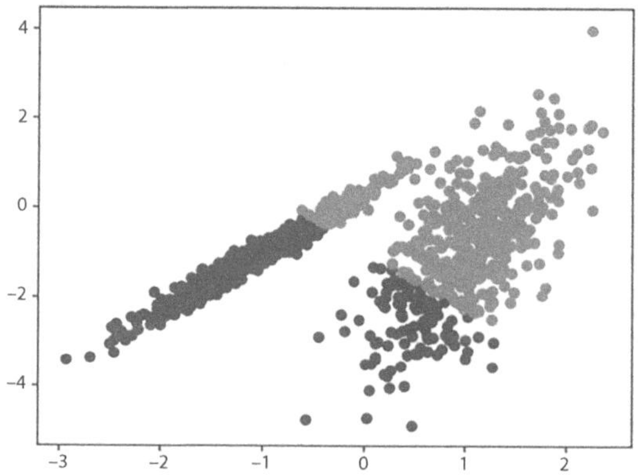

Figure 11.12 Scatter plot on K-means clustering.

```
from sklearn.cluster import Mini BatchMeans
from matplotlib import pyplot
# define dataset
X, = make_classification(n_samples-1000, n_features=2, n_informative-2,
n_redundant-0, n_clusters_per_class=1, random_state=4)
# define the model
model= MiniBatchKMeans (n_clusters-2)
# fit the model
model.fit(X)
# assign a cluster to each example
yhat= model.predict(X)
# retrieve unique clusters
clusters unique(yhat)
# create scatter plot for samples from each cluster
for cluster in clusters:
# get row indexes for samples with this cluster
row_ix where(yhat == cluster)
# create scatter of these samples
pyplot.scatter(x[row_ix, 0], x[row_ix, 1])
# show the plot
pyplot.show()
```

11.7.9 Mean Shift

Mean shift grouping involves finding and adjusting centroids basically dependent on the thickness of models inside the trademark space.

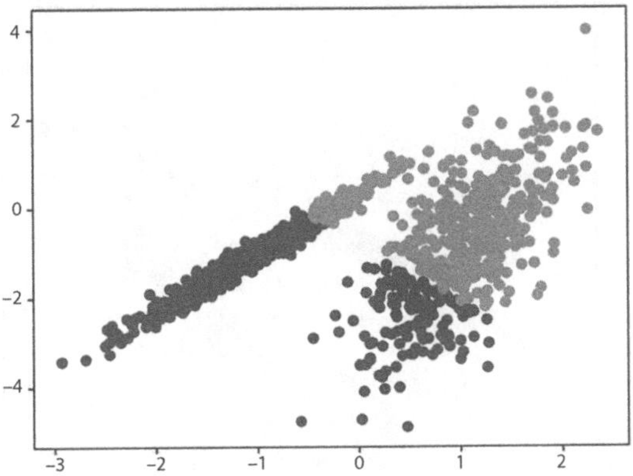

Figure 11.13 Scatter plot on mini-batch k-means.

The entire model is recorded under.

```
# mean shift clustering
from numpy import unique
from numpy import where
from sklearn.datasets import make_classification
from sklearn.cluster import MeanShift
from matplotlib import pyplot
# define dataset
X, = make_classification(n_samples-1000, n_features=2, n_informative=-2, n_redundant-0, n_clusters_per_class=1, random_state=4)
# define the model
model = MeanShift()
# fit model and predict clusters
yhat = model.fit_predict(X)
clusters= unique (yhat)
# create scatter plot for samples from each cluster
for cluster in clusters:
# get row indexes for samples with this cluster
row_ix= where(yhat cluster)
# create scatter of these samples
pyplot.scatter(x[row_ix, 0], x[row_ix, 1])
pyplot.show() # show the plot
```

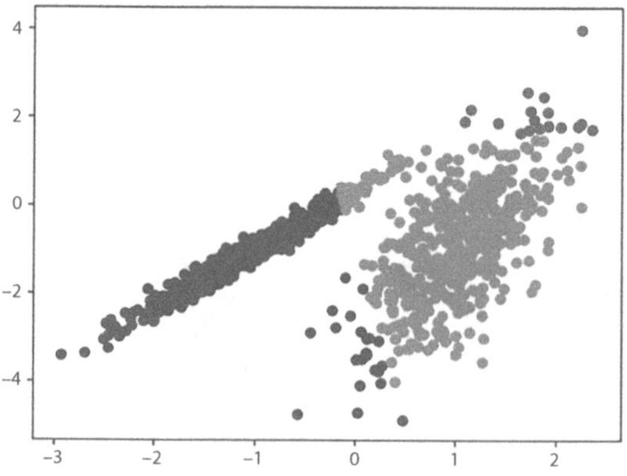

Figure 11.14 Scatter plot on mean shift clustering.

11.7.10 OPTICS

OPTICS grouping (wherein OPTICS is brief for ordering elements to see the clustering shape) is an adjusted model of DBSCAN characterized previously.

The total occurrence is recorded beneath.

```
# optics clustering
from numpy import unique
from numpy import where
from sklearn.datasets import make_classification
from sklearn.cluster import OPTICS
from matplotlib import pyplot
# define dataset
X, _ = make_classification(n_samples-1000, n_features=2, n_informa-
tive-2, n_redundant-0, n_clusters_per_class=1, random_state=4)
# define the model
model = OPTICS (eps-0.8, min_samples=10)
# fit model and predict clusters
yhat = model.fit_predict(X)
# retrieve unique clusters
clusters unique(yhat)
# create scatter plot for samples from each cluster
for cluster in clusters:
# get row indexes for samples with this cluster
row_ix = where(yhat cluster)
# create scatter of these samples
pyplot.scatter(x[row_ix, 0], X[row_ix, 1])
# show the plot
pyplot.show()
```

11.7.11 Unearthly or Spectral Clustering

Unearthly clustering is an in-vogue polish of grouping techniques, drawn from straight variable based math.

The entire model is filed underneath.

```
# spectral clustering
from numpy import unique
from numpy import where
from sklearn.datasets import make classification
```

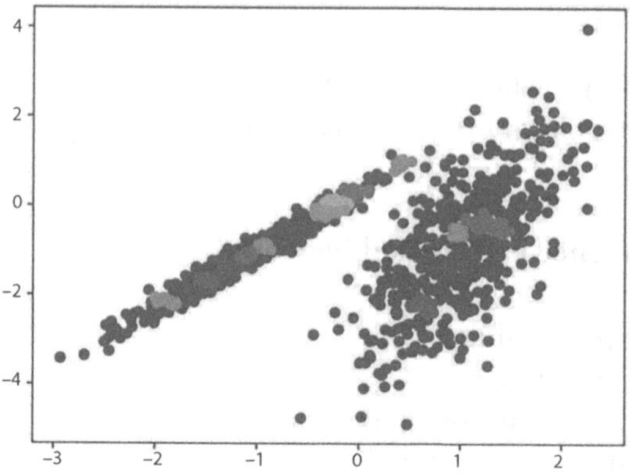

Figure 11.15 Scatter plot on OPTICS clustering.

```
from sklearn.cluster import SpectralClustering
from matplotlib import pyplot
# define dataset
X, = make_classification(n_samples-1000, n_features=2, n_informative-2,
n_redundant-0, n_clusters_per_class-1, random_state=4)
# define the model
model= SpectralClustering(n_clusters=2) =
# fit model and predict clusters
yhat= model.fit_predict(X)
# retrieve unique clusters
Clusters=unique(yhat)
# create scatter plot for samples from each cluster
for cluster in clusters:
# get row indexes for samples with this cluster
row_ix where(yhat == cluster)
# create scatter of these samples
pyplot.scatter(x[row_ix, 0], X[row_ix, 1])
pyplot.show() # show the plot
```

11.7.12 Gaussian Mixture Model

A Gaussian combination model sums up a multivariate chance thickness trademark with a combination of Gaussian freedom dispersions as its name indicates [6].

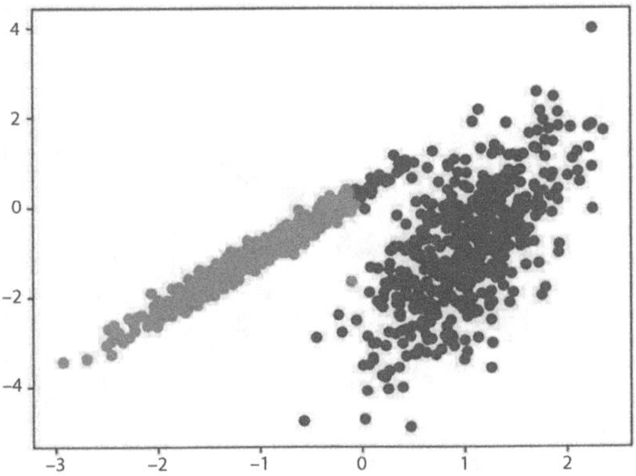

Figure 11.16 Scatter plot on spectral clustering.

The total example is filed under.

```
# gaussian mixture clustering
from numpy import unique
from numpy import where
from sklearn.datasets import make_classification
from sklearn.mixture import GaussianMixture
from matplotlib import pyplot
# define dataset
X, _ = make_classification(n_samples-1000, n_features=2, n_informative-2, n_redundant-0, n_clusters_per_class=1, random_state=4)
# define the model
model=GaussianMixture(n_components=2)
# fit the model
model.fit(X)
# assign a cluster to each example
yhat= model.predict(X)
# retrieve unique clusters
clusters= unique (yhat)
# create scatter plot for samples from each cluster
for cluster in clusters:
# get row indexes for samples with this cluster
row_ix = where(yhat == cluster)
# create scatter of these samples
pyplot.scatter(x[row_ix, 0], x[row_ix, 1])
# show the plot
pyplot.show()
```

[6]

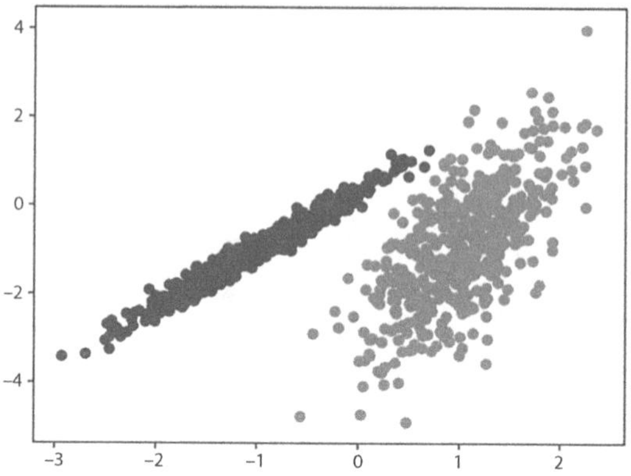

Figure 11.17 Scatter plot on gaussian mixture model.

11.8 Python Implementation of K-Means

Importing applicable libraries
import numpy as np
import pandas as pd
import statsmodels.api as sm
import matplotlin.pyplot as plt
import seaborn as sns
sns.set()
from sklearn.cluster import KMeans

11.8.1 Stacking the Data

data= pd.read_csv('Countryclusters.csv')
data

	Country	Latitude	Longitude	Language
0	USA	44.97	-103.77	English
1	Canada	62.40	-96.80	English
2	France	46.75	2.40	French
3	UK	54.01	-2.53	English
4	Germany	51.15	10.40	German
5	Australia	-25.45	133.11	English

Figure 11.18 Data samples.

11.8.2 Plotting the Information

```
plt.scatter(data['Logitude'],data['Latitude'])
plt.xlim(-180,180)
plt.ylim(-90,90)
plt.show()
```

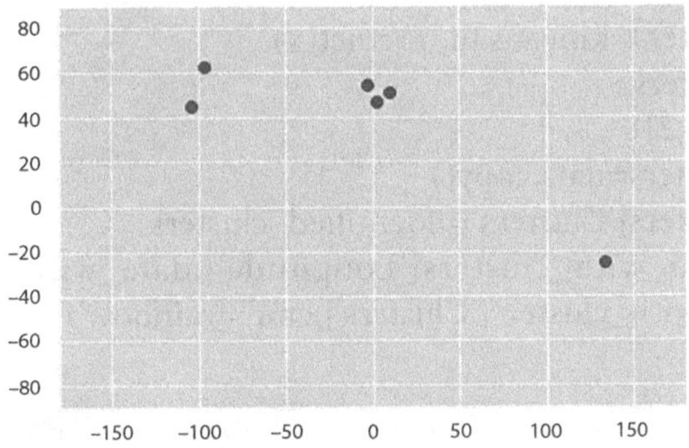

Figure 11.19 Scatter plot.

11.8.3 Choosing the Component

```
x=data.iloc[:,1:3] #lt for rows and second for columns
x
```

	Latitude	Longitude
0	44.97	-103.77
1	62.40	-96.80
2	46.75	2.40
3	54.01	-2.53
4	51.15	10.40
5	-25.45	133.11

Figure 11.20 Filtered data.

11.8.4 Clustering

```
Kmeans = KMeans(3)
Means.fit(x)
```

11.8.5 Clustering Results

```
Identified_clusters=kmeans.fit_predict(x)
Identified_clusters
array([1,1,0,0,0,2])
data_with_clusters=data.copy()
data_with_clusters['Clusters']=identified_clusters
plt.scatter(data_with_clusters['Longitude'],data_with_clusters['Latitude'],c=data_with_clusters['Clusters'],cmp='rainbow')
```

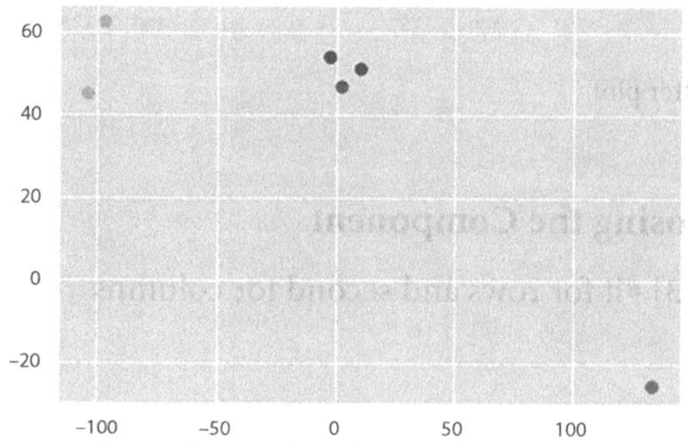

Figure 11.21 Scatter plot.

Endeavoring stand-out technique (to find no. of bunches to be settled on).

11.8.6 WCSS and Elbow Technique

```
wcss=[]
for i in range(1,7):
kmeans=KMean(i)
kmean.fit(x)
wcss_iter=kmeans.inertia_
wcss.append(wcss_iter)
number_clusters=range(1,7)
plt.plot(number_clusters,wcss)
```

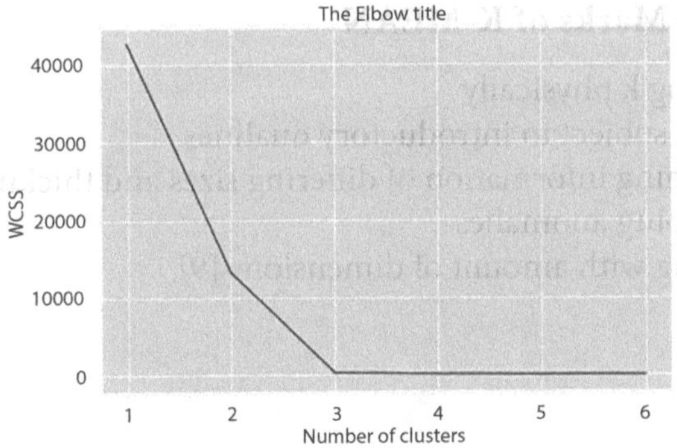

Figure 11.22 No. of clusters vs. WCSS.

plt.title('The Elbow title')
plt.xlabel('Number of clusters')
plt.ylabel('WCSS')

We can choose three as the number of groups; this methodology shows what the great number of clusters is [7].

11.8.7 Uses of K-Mean Bunching

K-mean grouping is the greatest typical unaided gadget dominating calculation. It is miles broadly utilized for parts programs which incorporate

1. Photo division
2. Consumer division
3. Species grouping
4. Anomaly identification
5. Clustering languages [8]

11.8.8 Benefits of K-Means

- Phenomenally simple to implement
- Scales to huge records units
- Ensures assembly
- Can warm-begin the places of centroids
- Easily adjusts to new models
- Sums up to groups of various sizes and styles, comprising of curved clusters [9]

11.8.9 Bad Marks of K-MEAN

- Picking k physically
- Being subject to introductory qualities
- Bunching information of differing sizes and thickness
- Grouping anomalies
- Scaling with amount of dimensions [9]

References

1. https://www.geeksforgeeks.org/clustering-in-machine-learning/
2. https://www.javatpoint.com/clustering-in-machine-learning
3. https://www.analyticsvidhya.com/blog/2016/11/an-introduction-to-clustering-and-different-methods-of-clustering/
4. https://developers.google.com/machine-learning/clustering/overview
5. https://towardsdatascience.com/the-5-clustering-algorithms-data-scientists-need-to-know-a36d136ef68
6. https://machinelearningmastery.com/clustering-algorithms-with-python/
7. https://www.analyticsvidhya.com/blog/2021/04/k-means-clustering-simplified-in-python/
8. https://www.kaggle.com/prashant111/k-means-clustering-with-python
9. https://developers.google.com/machine-learning/clustering/algorithm/advantages-disadvantages

12
Reinforcement Learning

Reinforcement learning (RL) is a kind of artificial intelligence (AI). It is far with regard to taking suitable development to boost acclaim in a specific situation. It is miles employed via different programming systems and machines to find the decent achievable conduct or course as it should absorb a chose circumstance. Reinforcement getting-to-know contrasts from managed learning in a manner that, in directed dominating, the training information has the arrangement key with it so the model is talented with the best response itself simultaneously, whereas, in support dominating, there is no response except for the support specialist that chooses what to do to play out the given mission. Without a trace of a tutoring dataset, it is miles sure to gain from its appreciate.

12.1 Model

The issue is as per the following: We have a specialist and an applause, with many obstacles. The specialist should find the outstanding possible way to arrive at the prize. The accompanying problem clarifies the difficulty all the more without any problem.

The above picture shows the robot, jewel, and chimney. The point of the computerized is to get the commendation that this is the valuable stone and to avoid the impediments that might be ended. The automated learns by utilizing endeavoring each of the suitable ways after which picking the way that rewards him with the least obstacles. Each appropriate advance will give the robot a recognition, and each off-base advance will take away the commendation of the automated. The whole recognition could be determined while it arrives at the absolute last prize that is the precious stone [1].

Kolla Bhanu Prakash. Machine Learning for Industrial Applications, (205–224) © 2024 Scrivener Publishing LLC

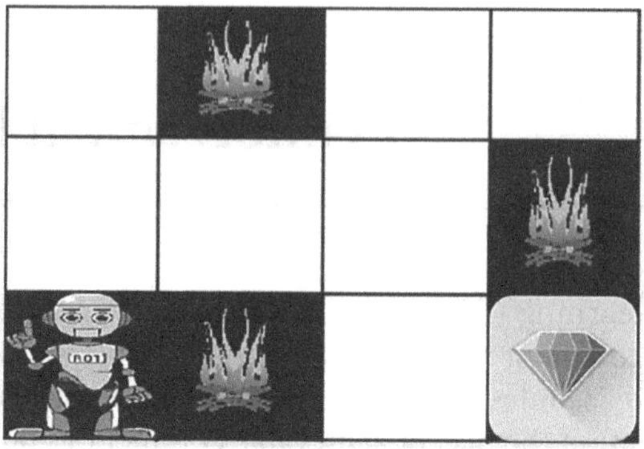

Figure 12.1 An example for reinforcement learning.

12.2 Terms Utilized in Reinforcement Learning

- **Specialist():**
  An element that can comprehend/investigate the environmental factors and follow up on it.
- **Climate():**
  A situation wherein a specialist is available or encircled through. In RL, we depend on the stochastic climate; this implies that it is arbitrary.
- **Activity():**
  Moves are the activities taken through a specialist inside the environmental factors.
- **State():**
  State is a situation back through the environmental elements after each movement taken by the specialist.
- **Reward():**
  An input got back to the specialist from the environmental factors to assess the movement of the specialist.
- **Strategy():**
  Inclusion is a strategy applied through the specialist for the following movement fundamentally based at the contemporary country.
- **Regard():**
  It is normal postponed term retuned with the cut worth thing and instead of the passing prize.

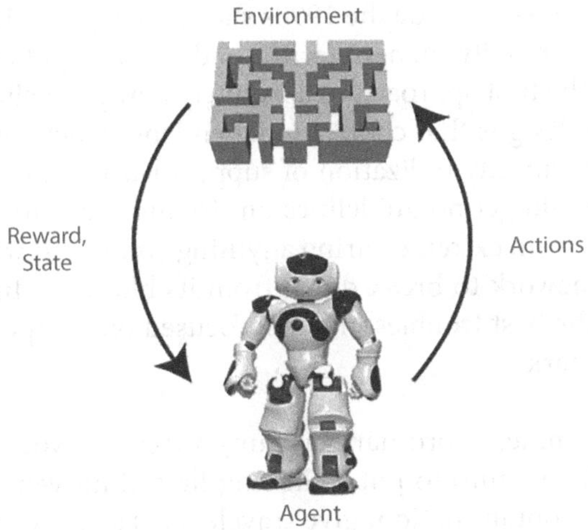

Figure 12.2 Interaction between the agent and the environment.

- **Q-Value():**
  It is far generally comparable as the value, yet it acknowledges an additional one limit as a contemporary turn of events [2].

12.3 Key Elements of Reinforcement Learning

- In RL, the specialist is not constantly told with regard to the environmental factors and what moves should be made.
- It is far reliant upon fundamental cycle.
- The specialist makes the accompanying move and changes states in accordance with the remarks of the past development.
- The specialist can likewise become bogged down acclaim.
- The environmental factors are stochastic, and the specialist needs to find it to accomplish the most extreme great prizes [2].

12.4 Instances of Reinforcement Learning

Projects of support considering had been inside the past restricted through frail PC framework. In any case, as Gerard Tesauro's backgammon AI superplayer progressed inside the 1990s shows a progress. That early headway is, at present, rapidly changing with solid new computational advancements starting the best approach to absolutely new propelling ventures.

Training the designs that control independent vehicles is an incredible occurrence of a capacity utilization of support learning. In an ideal situation, the PC should get no guidelines on driving the auto. The developer could stay away from extreme wiring anything connected with the test and permit the framework to break down from its blunders. In an incredible circumstance, the best troublesome that focused on component would be the prize trademark.

- For example, in ordinary circumstances, we would require a confident auto to put wellbeing first, limit venture time, reduce contamination, give travelers solace, and submit to the arrangements of guideline. The software engineer cannot foresee the entire parcel that would show up in the city. In tendency to creating postponed "expecting to be then" rules, the engineer prepares the support learning specialist to have the option to dominating from the contraption of remunerations and results. The specialist (another name for support acquiring information on calculations playing out the mission) gets awards for arriving at exceptional longings.
- Each and every other model: Deepsense.ai partook inside the "figuring out how to run" mission, which meant to teach an advanced sprinter without any preparation. The sprinter is a refined and specific outer muscle variant planned through the Stanford Neuromuscular Biomechanics Laboratory. Concentrating on the specialist how to run is an essential advance in building a fresh out of the plastic new age of prosthetic legs, ones that precisely perceive people by walking styles and change themselves to simplify moving and all the more impressive. Indeed, even as it is possible and has been done in Stanford's labs, extreme wiring the orders as a whole and anticipating all feasible styles of walking require heaps of canvases from observably proficient developers [3].

12.5 Advantages of Reinforcement Learning

Support dominating applies to a broad scope of intricate issues that cannot be handled with various framework ruling computations. RL is toward fake general information (AGI), as it can search for a lengthy term objective while examining diverse possible results independently. Some of the upsides of RL incorporate the following:

- **Has some expertise in the issue as a whole:** Conventional device acquiring information on calculations are intended to dominate at specific subtasks, without a thought of the enormous photograph. RL, be that as it may, does not separate the issue into subproblems; it promptly attempts to augment the drawn out remuneration. It has a conspicuous explanation, knows the goal, and can trade off momentary awards for long-term benefits.
- **Does now not need a different measurements assortment step:** In RL, preparing realities is gotten through the immediate association of the specialist with the environmental factors. Tutoring insights is the concentrating on specialist's revel in, as of now, not a different assortment of measurements that should be taken care of to the calculation. This fundamentally decreases the weight at the director in cost of the tutoring system.
- **Works in powerful, questionable conditions:** RL calculations are intrinsically versatile and built to answer to changes inside the environmental elements. In RL, time matters, and the partake in that the specialist gathers is not autonomously and indistinguishably disseminated, dislike customary gadget dominating calculations. For the explanation that element of time is profoundly covered inside the mechanics of RL, learning is innately versatile [4].

12.6 Challenges with Reinforcement Learning

Indeed, even as RL calculations have been an achievement in fixing jumbled issues in different reproduced conditions, their gathering truly has been slow. Here are a portion of the requesting circumstances that have made their take-up extreme.

- **RL specialist wants standard size revel in:** RL techniques independently create tutoring data through collaborating with the environmental elements. As is it needed, the charge of records series is limited via the elements of the environmental elements. Conditions with unreasonable dormancy delayed down the dominating bend. In addition, in confounded conditions with extreme layered country spaces, huge investigation is required before an astonishing arrangement can be found.
- **Delayed prizes:** The dominating specialist can substitute off brief time frame period compensations for long-lasting period gains. While this primary rule makes RL helpful, it, likewise, makes it intense for the specialist to find the most valuable inclusion. This is extraordinarily legitimate in conditions in which the result is obscure till a tremendous assortment of consecutive moves are taken. On this situation, doling out credit to a former movement for the result is hard and can present large difference all through schooling. The game of chess is a significant occasion here, in which the result of the game is obscure till every gamer has made every one of their developments.
- **Absence of interpretability:** When a RL specialist has discovered the surest strategy and is sent inside the climate, it makes moves basically dependent on its appreciate. To an outer onlooker, the justification behind those activities will not be self-evident. This deficiency of interpretability obstructs the improvement of acknowledge as obvious with between the specialist and the spectator. Assuming a spectator ought to clarify the developments that the RL specialist obligations, it may help him in issuing the data better and finding obstructions of the variant, extraordinarily in high-chance conditions [4].

12.7 Sorts of Reinforcement

There are two styles of reinforcement, as follows.

12.7.1 Positive

Encouraging feedback is characterized as, while an event happens due to a particular conduct, it builds the power and the recurrence of the direct. In various expressions, it positively affects direct.

Advantages of Reinforcement Learning:

- ➢ Maximizes generally speaking execution;
- ➢ Preserves trade for a drawn out length;
- ➢ Too much reinforcement can prompt an over-burden of states that could diminish the outcomes.

12.7.2 Negative

Helpless reinforcement is characterized as fortifying of lead because of the reality that an awful circumstance is halted or then again redirected.

Advantages of Reinforcement Learning:

- ➢ Builds lead;
- ➢ Offers protection from a unimportant standard of as a rule execution;
- ➢ It least difficult presents to the point of meeting up the insignificant conduct [1].

12.8 What are the Useful Utilizations of Reinforcement Learning?

Taking into account that RL requires various information, in this manner, it is miles greatest material in spaces wherein data are mimicked and a popular issue like interactivity and advanced mechanics are not utilized.

- RL is broadly utilized in building AI for playing PC computer games. AlphaGo 0 is the essential PC programming to overcome a worldwide boss inside the antiquated Chinese language round of pass. Others comprise ATARI computer games, Backgammon, and numerous others.
- In advanced mechanics and business mechanization, RL is utilized to permit the robot to make a green versatile control

> machine for itself that gains from its own revel in and direct. DeepMind's artwork on deep reinforcement acquiring information for mechanical manipulation with asynchronous inclusion refreshes is a decent example of the indistinguishable. Watch this thrilling show video.

Various projects of RL comprise abstractive literary substance synopsis motors, discussion marketers (textual content and discourse), which can study from individual collaborations and improve with time, concentrating on most positive cure rules in medical care, and RL-based sellers for online stock exchanging [6].

12.9 How is Reinforcement Learning Extraordinary from Supervised and Unsupervised Learning?

Support contemplating is viewed as its part of gadget examining, despite the fact that it has a couple of likenesses to different styles of contraption getting to know, which is wrecked down into the ensuing four space names.

12.9.1 Supervised Learning

In directed getting to know, calculations instruct on an edge of arranged realities. Directed learning calculations can least complex examine credits that are assigned inside the records set. Normal projects of administered considering are picture notoriety models. These designs get an immovable of arranged pictures and look at to separate normal ascribes of predefined organization.

12.9.2 Unsupervised Learning

In unaided dominating, developers flip calculations free on completely unlabeled insights. The arrangement of rules learns by indexing its perceptions about record capacities without being determined what to search for.

12.9.3 Semisupervised Learning

This strategy takes a middle floor method. Developers enter a phenomenally little arrangement of characterized preparing insights, just as a greater corpus of unlabeled data. The arrangement of rules is then educated to extrapolate what it gains from the classified information to the unlabeled records and reach determinations from the set all in all.

12.9.4 Reinforcement Learning

This takes a particular technique through and through. It arranges a specialist in environmental elements with clean boundaries characterizing helpful interest and nonbeneficial interest and an all-encompassing final stage to reach. It is comparable, to a great extent, to directed securing data in that fashioners should give computations explicit dreams and describe prizes and teachings. This shows that the degree of explicit programming required is more than in unaided learning. However, when these boundaries are set, the calculation works on its own, making it tons more noteworthy self-coordinated than administered dominating calculations. Along these lines, people occasionally talk over with support contemplating as a part of semi-coordinated learning, but without a doubt, it is most commonly communicated as its sort of contraption learning [5].

12.10 Ways to Deal with Execute Reinforcement Learning

There are three methods for executing support learning in ML, which may be as follows.

12.10.1 Value-Based

The expense-based absolutely technique is prepared finding the ideal value included, which is the most extreme worth at a realm under any arrangement. Subsequently, the specialist expects the extended term return at any nation(s) underneath strategy π.

12.10.2 Policy-Based

A methodology-based technique is to track down the most fitting consideration for the most likely remunerations without the usage of the value part. In this technique, the specialist attempts to utilize this sort of inclusion that the movement finished in each progression assists with augmenting the future recognition.

The strategy-based methodology has especially two assortments of inclusion:

- **Deterministic:** The equivalent development is created by the inclusion (π) at any state.
- **Stochastic:** In this inclusion, chance decides the delivered activity.

12.10.3 Model-Based

Inside the model-based technique, an advanced variant is made for the environmental elements, and the specialist investigates that climate to learn it. There might be no particular response or set of rules for this procedure because of the reality that the form portrayal is exceptional for every climate [2].

12.11 Components of Reinforcement Learning

There are four principle components of RL, which are given beneath:

1. Policy
2. Reward signal
3. Value function
4. Model of the climate

12.11.1 Policy

An arrangement might be characterized as a way how a specialist acts at a given time. It maps the apparent conditions of the climate to the moves taken on those states. An approach is the center component of the RL as it alone can diagram the conduct of the specialist. In a couple of cases, it very well may be a simple element or an exploration table, whereas, for different examples, it might contain famous calculation as a look for method. It can be deterministic or a stochastic strategy:

For deterministic arrangement: $\mathbf{x = \pi(a)}$
For stochastic arrangement: $\mathbf{\pi(x \mid a) = P[At = x \mid St = a]}$

12.11.2 Reward Signal

The reason for support getting-to-know is depicted through the commendation sign. At every country, the environmental elements send an immediately sign to the getting-to-know specialist, and this sign is alluded to as an award sign. These prizes are given in accordance with the best and horrendous moves taken through the specialist. The specialist's central goal is to expand the whole number of remunerations for positive developments. The award sign can change the inclusion, for example; however, a movement settled on via the specialist brings about a low commendation; then, at that point, the inclusion can, likewise, exchange to pick different developments inside the fate.

12.11.3 Value Function

The charge gives realities around how precise the situation and movement are and what kind of recognition a specialist can anticipate. An applause recommends the quick sign for each careful and awful activity, whereas a charge work indicates the coolest country and activity for what's to come. The value limit relies upon the commendation as, without compensation, there might be no expense. The reason for assessing values is to accomplish more rewards.

12.11.4 Model

The excess detail of support acquiring information is the model that copies the conduct of the environmental elements. With the assistance of the form, you could make inductions concerning how the climate will act. Alongside, assuming a state and an activity are given, then, at that point, a model can are anticipating the following realm and prize.

The model is utilized for making arrangements, which implies that it gives a way to take a course of movement through pondering all future conditions sooner than practically encountering the ones conditions. The methodologies for fixing the RL issues with the assistance of the rendition are named the model-based absolutely technique. Similarly, a methodology without the use of a form is known as a model-free strategy [2].

12.12 Typical Reinforcement Learning Algorithms

Rather than interfacing with a specific estimation, the circle of support getting-to-know is composed of a few calculations that take really exceptional systems. The distinctions are extraordinarily a result of their methods for investigating their surroundings.

12.12.1 State-Activity Reward-State-Activity (SARSA)

This support learning calculation is developed through giving the specialist what is known as an approach. The inclusion is basically a possibility that tells it the rates of positive moves, resulting in remunerations or valuable states.

12.12.2 Q-Learning

This procedure to support getting-to-know adopts the other strategy. The specialist gets no approach, which means that its investigation of its environmental elements is additional self-coordinated.

12.12.3 Profound Q-Networks

Those calculations use neural organizations correspondingly to support concentrating on methodologies. They utilize oneself coordinated environmental factors investigation of support dominating. Predetermination moves depend absolutely on an arbitrary example of past helpful activities found through the neural organization [5].

12.13 Support Learning Algorithm: Python Implementation Utilizing Q-Learning

In this, we will exhibit how to carry out an essential RL calculation, which is known as the Q-learning procedure. In this, we endeavor to encourage a bot to arrive at its objective utilizing the Q-learning procedure.

Stage 1: Importing the necessary libraries

```
Import numpy as np
Import pylab as pl
Import networkx as nx
```

Stage 2: Defining and envisioning the diagram

```
Edges = [(0, 1), (1, 5), (5, 6), (5, 4), (1, 2), (1, 3), (9, 10), (2, 4), (0, 6), (6, 7),
(8, 9), (7, 8), (1, 7), (3, 9)]
goal = 10
G = nx.Graph()
G.add_edges_from (edges)
Pos= nx.spring_layout(G)
nx.draw_networkx_nodes (G, pos)
nx.draw_networkx_edges (G, pos)
nx.draw_networkx_labels (G, pos)
pl.show()
```

Stage 3: Defining the award the framework for the bot

```
MATRIX_SIZE = 11
M = np.matrix(np.ones(shape=(MATRIX_SIZE, MATRIX_SIZE)))
M *= -1
for point in edges:
   Print(point)
   If point [1]== goal:
      M[point] = 100
   else:
      M[point] = 0
   If point[0]== goal:
      M[point[::-1]]= 100
   else:
      M[point[::-1]]= 0 # reverse of point
M[goal, goal]=100
```

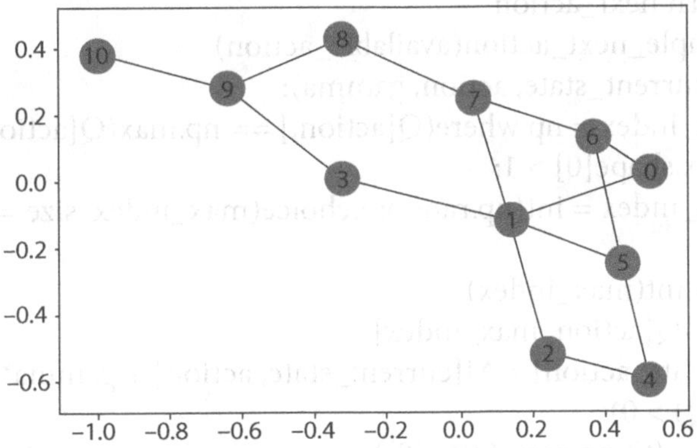

Figure 12.3 Graph plot.

```
[[ -1.   0.  -1.  -1.  -1.  -1.   0.  -1.  -1.  -1.  -1.]
 [  0.  -1.   0.   0.  -1.   0.  -1.   0.  -1.  -1.  -1.]
 [ -1.   0.  -1.  -1.   0.  -1.  -1.  -1.  -1.  -1.  -1.]
 [ -1.   0.  -1.  -1.  -1.  -1.  -1.  -1.  -1.   0.  -1.]
 [ -1.  -1.   0.  -1.  -1.   0.  -1.  -1.  -1.  -1.  -1.]
 [ -1.   0.  -1.  -1.   0.  -1.   0.  -1.  -1.  -1.  -1.]
 [  0.  -1.  -1.  -1.  -1.   0.  -1.   0.  -1.  -1.  -1.]
 [ -1.   0.  -1.  -1.  -1.  -1.   0.  -1.   0.  -1.  -1.]
 [ -1.  -1.  -1.  -1.  -1.  -1.  -1.   0.  -1.   0.  -1.]
 [ -1.  -1.  -1.   0.  -1.  -1.  -1.  -1.   0.  -1. 100.]
 [ -1.  -1.  -1.  -1.  -1.  -1.  -1.  -1.  -1.   0. 100.]]
```

Figure 12.4 Matrix M.

```
Print (M)
# add goal point round trip
```

Stage 4: Defining a few utility capacities to be utilized in the preparation

```
Q = np.matrix(np.zeros([MATRIX SIZE, MATRIX_SIZE]))
gamma = 8.75
# learning parameter
Initial_state= 1
# Determines the available actions for a given state
def available_actions(state):
        current_state_row= M[state, ]
        available_action = np.where(current_state_row>= 0) [1]
        return available_action
available_action = available_actions (initial_state)
#Chooses one of the available actions at random
def sample_next_action(available_actions_range):
        next_action = int(np.random.choice(available_action, 1))
        return next_action
action = sample_next_action(available_action)
def update(current_state, action, gamma):
        max_index = np.where(Q[action,] == np.max(Q[action, ]))[1]
If max_index.shape[0] > 1:
        max_index = int(np.random.choice(max_index, size = 1))
else:
max_index= int(max_index)
max_value = Q[action, max_index]
Q[current_state, action] = M[current_state, action] + gamma*max_value
If (np.max(Q) > 0):
return(np.sum(Q/np.max(Q)*100))
else:
```

```
return (0)
# Updates the Q-Matrix according to the path chosen
update(initial_state, action, gamma)
```

Stage 5: Training and assessing the bot utilizing the Q-matrix

```
Scores = []
for i in range(1000):
Current_state= np.random.randint(0, int(Q.shape[0]))
available_action = available_actions (current_state)
action = sample_next_action(available_action)
score = update (current_state, action, gamma)
Scores.append(score)

# print("Trained Q matrix:")

# print(Q / mp.max(0)*100) You can uncomment the above two lines to view the trained Q matrix

#Testing

Current_state=0

Steps=[current_state]

While Current_state!= 10:
next_step_index = np.where(Q[current_state, ] == np.max(Q[current_st

If next_step_index.shape[0] > 1:
next_step_index = int(np.random.choice (next_step_index, size =1
else:
```

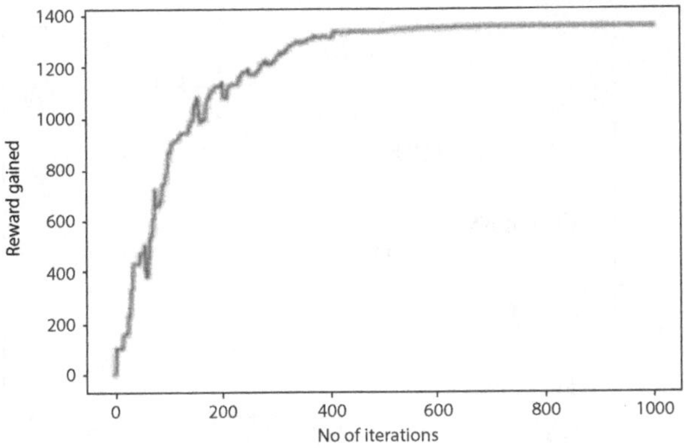

Figure 12.5 Numbers of iterations vs. rewards gained.

```
        next_step_index = int(next_step_index)
steps.append(next_step_index)

Current_state= next_step_index
Print("Most efficient path:")
Print (steps)
Pl.plot(scores)
Pl.xlabel('No of iterations')
Pl.ylabel('Reward gained')
Pl.show()
```

Most efficient path: [0, 1, 3, 9, 10]

Presently, permit conveys this bot to a more sensible putting, allowing us to envision that the bot is an analyst and is looking for out the area of an enormous medication racket. He normally reasons that the street pharmacists would not sell their product in a space; this is recognized to be regularly visited by utilizing the police, and the elevating places are near the spot of the medication racket. Moreover, the dealers leave a follow in their product wherein they sell them, and this might help the investigator in discovering the predefined place. We need to prepare our bot to find the district the use of those environmental clues [7].

Stage 6: Defining and envisioning the new diagram with the ecological pieces of information

```
#Defining the locations of the police and the drug traces
Police = [2, 4, 5]
drug_traces= [3, 8, 9]
G= nx.Graph()
```

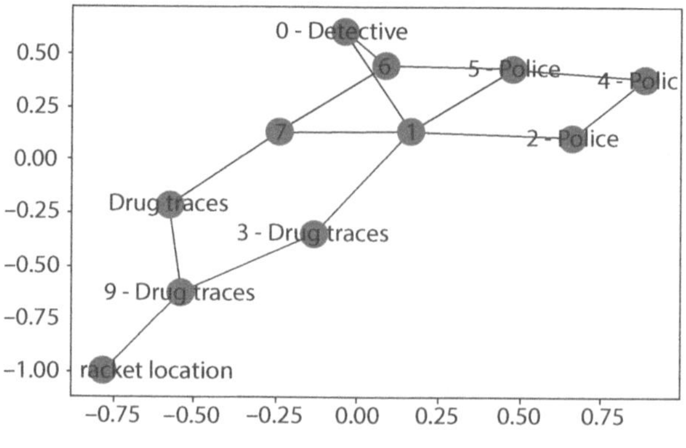

Figure 12.6 Graph with the ecological pieces of information.

```
G.add_edges_from(edges)
mapping = {0:'0- Detective', 1:'1', 2:'2 – Police', 3:'3 – Drug traces'
        4:'4 -Police', 5:'5 Police', 6: '6', 7:'7', 8: 'Drug traces'
        9:'9 -Drug traces', 10: '10- Drug racket location'}
H= nx.relabel_nodes (G, mapping)
Pos= nx.spring_layout (H)
nx.draw_networkx_nodes (H, pos, node_size =[200, 200, 200, 200, 200,
200,
nx.draw_networkx_edges (H, pos)
nx.draw_networkx_labels (H, pos)
pl.show()
```

Note:
The above diagram may, likewise, look somewhat particular from the first chart; however, they are the equivalent diagrams. This is because of the arbitrary arrangement of hubs through the networkx library.

Stage 7: Defining a few utility capacities for the preparation interaction

```
Q = np.matrix(np.zeros([MATRIX_SIZE, MATRIX_SIZE]))
env_police = np.matrix(np.zeros([MATRIX_SIZE, MATRIX _SIZE]))
env_drugs = np.matrix(np.zeros([MATRIX_SIZE, MATRIX_SIZE]))
initial_state= 1
#5ame as above
def available_actions(state):
        Current_state_row = M[state, ]
        av_action = np.where(current_state_row >= 0)[1]
return av_action
def sample_next_action(available_actions_range):
        next_action = int(np.random.choice(available_action, 1))
        return next_action
#Same as above
#Exploring the environment
def collect_environmental_data (action):
        found = []
if action in police:
        found.append('p')
If action in drug traces:
        found.append("d")
return (found)
available_action = available_actions (initial_state)
action=sample_next_action(available_action)
```

```
def update(current_state, action, gamma):
        max_index = np.where(Q[action,] == np.max(Q[action, ]))[1]
If max_index.shape[0] > 1:
        max_index= int(np.random.choice(max_index, size= 1))
else:
        max_index = int(max_index)
max_value = Q[action, max_index]
Q[current_state, action] = M[current_state, action] +gamma*max_value
Environment= collect_environmental_data(action)
If 'p' in environment:
        env_police[current_state, action] += 1
If 'd' in environment:
        env_drugs[current state, action] += 1
If (np.max(Q) > 0):
        return(np.sum(Q/np.max(Q)*100))
else:
        return (0)
#Same as above
Update(initial_state, action, gamma)
def available_actions_with_env_help(state):
current_state_row= M[state, ]
av_action = np.where(current_state_row >= 0)[1]
# if there are multiple routes, dis-favor anything negative
env_pos_row= env_matrix_snap[state, av_action]
If (np.sum(env_pos_row <0)):
# can we remove The negative directions from av_act?
temp_av_action = av_action[np.array(env_pos_row)[0]>=0]
if len(temp_av_action) > 0:
        av_action = temp_av_action
return av_action
#Determines the available actions according to the environment
```

Stage 8: Visualizing the environmental lattices

```
Scores = []
For i in range(1000):
        Current_state= np.random.randint(0, int(Q.shape[0]))
        available_action = available_actions (current_state)
        action = sample_next_action(available_action)
        score=update (current_state, action, gamma)
#print environmental matrices
Print('Police Found')
```

```
Police Found
[[ 0.  0.  0.  0.  0.  0.  0.  0.  0.  0.  0.]
 [ 0.  0. 23.  0.  0. 14.  0.  0.  0.  0.  0.]
 [ 0.  0.  0.  0. 51.  0.  0.  0.  0.  0.  0.]
 [ 0.  0.  0.  0.  0.  0.  0.  0.  0.  0.  0.]
 [ 0.  0. 51.  0.  0. 37.  0.  0.  0.  0.  0.]
 [ 0.  0.  0.  0. 29.  0.  0.  0.  0.  0.  0.]
 [ 0.  0.  0.  0.  0. 32.  0.  0.  0.  0.  0.]
 [ 0.  0.  0.  0.  0.  0.  0.  0.  0.  0.  0.]
 [ 0.  0.  0.  0.  0.  0.  0.  0.  0.  0.  0.]
 [ 0.  0.  0.  0.  0.  0.  0.  0.  0.  0.  0.]
 [ 0.  0.  0.  0.  0.  0.  0.  0.  0.  0.  0.]]
```

Figure 12.7 Police found matrix.

```
Drug traces Found
[[ 0.  0.  0.  0.  0.  0.  0.  0.  0.  0.  0.]
 [ 0.  0.  0. 12.  0.  0.  0.  0.  0.  0.  0.]
 [ 0.  0.  0.  0.  0.  0.  0.  0.  0.  0.  0.]
 [ 0.  0.  0.  0.  0.  0.  0.  0.  0. 40.  0.]
 [ 0.  0.  0.  0.  0.  0.  0.  0.  0.  0.  0.]
 [ 0.  0.  0.  0.  0.  0.  0.  0.  0.  0.  0.]
 [ 0.  0.  0.  0.  0.  0.  0.  0.  0.  0.  0.]
 [ 0.  0.  0.  0.  0.  0.  0.  0. 29.  0.  0.]
 [ 0.  0.  0.  0.  0.  0.  0.  0.  0. 52.  0.]
 [ 0.  0.  0. 36.  0.  0.  0.  0. 37.  0.  0.]
 [ 0.  0.  0.  0.  0.  0.  0.  0.  0. 47.  0.]]
```

Figure 12.8 Drug-traced found matrix.

Print(env_police)
Print(' ')
Print('Drug traces Found')
Print (env_drugs)

Stage 9: Training and assessing the model

Scores = []
for i in range(1000):
Current_state= np.random.randint(0, int(Q.shape[0]))
available_action = available_actions_with_env_help (current_state)
action = sample_next_action(available_action)
Score= update (current_state, action, gamma)
Scores.append(score)
Pl.plot(scores)
Pl.xlabel('Number of iterations')
Pl.ylabel('Reward gained')
Pl.show()
[7]

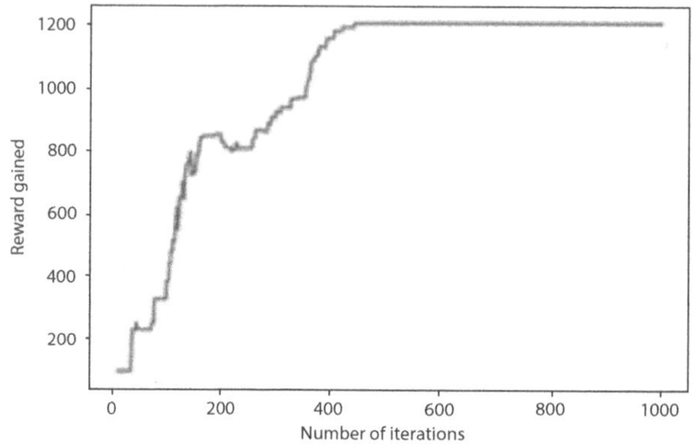

Figure 12.9 Rewards gained vs. numbers of iterations.

References

1. https://www.geeksforgeeks.org/what-is-reinforcement-learning/
2. https://www.javatpoint.com/reinforcement-learning
3. https://deepsense.ai/what-is-reinforcement-learning-the-complete-guide/
4. https://www.synopsys.com/ai/what-is-reinforcement-learning.html
5. https://www.techtarget.com/searchenterpriseai/definition/reinforcement-learning
6. https://towardsdatascience.com/reinforcement-learning-101-e24b50e1d292
7. https://www.geeksforgeeks.org/ml-reinforcement-learning-algorithm-python-implementation-using-q-learning/

13
Recommender Systems

13.1 Recommender Frameworks

Recommender frameworks are the constructions that can be intended to prescribe matters to the client basically dependent on various elements. Those constructions are expecting the most probable item that the clients are most potentially to buy and are important to them. Bunches like Netflix and Amazon use recommender constructions to help their clients to see the right item or films for them.

The recommender contraption manages a major degree of insight by utilizing and separating the most extreme basic records based absolutely on the data outfitted through a purchaser and various elements that are arranged with the customer's inclination and leisure activity. It uncovers the shape among buyer and object and attributes the likenesses among clients and articles for counsel.

Each of the clients and the contributions have profited from these styles of frameworks. The quality and decision-making strategy have, furthermore, progressed through those sorts of structures [1].

13.1.1 It Has the Accompanying Advantages

- Clients benefit from having the option to observe objects that have side interest.
- It helps organizations in getting their items to the right people.
- Clients may be fit to see items that are most extreme pertinent to them.
- It has content that is custom-made to the person.
- It helps destinations in creating individual responsibility [7].

Kolla Bhanu Prakash. Machine Learning for Industrial Applications, (225–244) © 2024 Scrivener Publishing LLC

13.1.2 What Defines an Extraordinary Suggestion?

Sorting out what characterizes a generally excellent suggestion is an issue in itself that many organizations fight with. This meaning of "attractive" hints looks at the general exhibition of the recommender that you developed. The top notch of a proposal might be surveyed through various strategies that measure inclusion and precision. Exactness is the negligible portion of precise tips out of absolute possible clues simultaneously as protection estimates the small part of articles inside the pursuit space that the contraption can offer clues for. The method of appraisal of a suggestion is absolutely reliant upon the dataset and strategy used to produce the proposal. Recommender structures extend various calculated likenesses with the class and relapse displaying inconvenience. In a truly ideal situation, you would have to perceive how real clients respond to suggestions and tune measurements across the individual to further develop your proposal; nonetheless, this is really difficult to achieve. To be expected, factual exactness measures to assess the precision of a recommender are root meansquare deviation (RMSD), mean absolute error (MAE), and k overlap go-approval.

13.1.2.1 K-Fold Cross-Validation

- Consider you have developed an adaptation so as to anticipate how well a client will value a thing essentially dependent on a rigid of capacities. Good overlay pass-endorsement may be used to infer the aftereffects of the model through precision estimations.
- Vague idea as a show test cut up, other than we make alright many haphazardly appointed training and actually look at sets.
- Each preparation set/overlay is utilized to prepare on the guidance gadget freely, after which the degree of the exactness of the subsequent frameworks is investigated against the set.
- We take the normal exactness rating to look how appropriately the proposal machine is contemplating.
- This methodology is gainful to keep your variant from overfitting, and, at the same time, it is a computationally huge cycle.

13.1.2.2 MAE (Mean Absolute Error)

$$\text{MAE} = \frac{\sum_{i=1}^{n} |y_i - x_i|}{n}$$

MAE = mean absolute error
y_i = prediction
x_i = true value
n = total number of data points

- This addresses the normal outright charge of every mistake in score expectation.
- The lower the MAE score, the better.

13.1.2.3 RMSD (Root Mean Square Deviation)

$$\text{RMSD} = \sqrt{\frac{\sum_{i=1}^{N} (x_i - \hat{x}_i)^2}{N}}$$

RMSD = root-mean-square deviation
i = variable i
N = number of non-missing data points
x_i = actual observations time series
$\hat{x}_i$ = estimated time series

- A comparable measurement to MAE, anyway, has a more grounded punishment for when the expectation might be extremely distant from the real expense and a more fragile punishment for when the forecast is toward the appropriate expense.
- It takes the squares off the qualification of legitimate and anticipated qualities in inclination to the amount of the outright qualities. This ensures that the subsequent expense is typically fine and is bigger when the thing that matters is high and more modest while the qualification is low.

The decline in the RMSD score, the better; those measurements are frequently used to evaluate the good suggestion; anyway, they need differently added substances. Having shopper data connected with hints is vital to perceive the genuine top notch of a proposal. Being equipped for track hit costs of suggestions, commitment to the stage, responsiveness, and numerous others will give a more clear angle of proposal uncommon. Various parts to be aware of are to know when to trade rules while the customer has not participated with them for a X measure of time or when to re-train recommenders dependent on new evaluations or associations from the clients. You would, moreover, have to observe whether or not those tips are limiting the client to a subsection of the items and how does the recommender adapt to curiosity, reach, and determination inclination. A/B giving a shot is regularly the procedure used to keep up with tune of those measurements (investigate our articles on Bayesian and frequentist a/b testing) [5].

13.1.3 What Can Be Suggested?

Numerous different things can be upheld through the contraption like movies, books, data, articles, occupations, and ads. Netflix utilizes a recommender device to prescribe films and web assortment to its clients. Further, YouTube suggests particular movies. There are various instances of recommender structures that are broadly utilized nowadays [1].

13.1.4 For What Reason Will We Really Want Recommender Frameworks?

Organizations utilize recommender structures that considers expanding pay on account of exceptionally modified offers and a better supporter appreciation.

Rules typically speed up searches and make it less convoluted for clients to get admission to the substance material that they are intrigued by and wonder them with things that they might have never searched for.

Furthermore, associations can benefit and keep up with customers via conveying messages with connections to new offers that meet the beneficiaries' hobbies or pointers of motion pictures and tv demonstrating their profiles.

The client starts to encounter perceived and comprehended items and is bound to purchase additional items or gobble up more substance. Through knowing what a client needs, the association benefits a cutthroat increase, and the danger of losing a buyer to a contender diminishes.

Given the acquainted expense of the clients, the guide comprising tips in frameworks and products is appealing. Additionally, it allows the enterprises to put in advance in their rivals and, in the end, blast their profits [3].

13.2 Use-Instances of Suggestion Framework

There are many use instances of the framework. A couple are as follows.

A. Customized content:
It works to improve the on-site insight via developing unique proposals for unmistakable kinds of crowds like netflix does.

B. Better item search knowledge:
It engages to class the thing subject to their abilities, e.g., texture and season [2].

13.3 Phases of Proposal Technique

13.3.1 Reality Series Fragment

This gathers pertinent data of clients to produce a client profile or model for the expectation errands alongside the individual's characteristic, practices, or content of the resources the shopper gets to. A proposal specialist cannot include accurately till the purchaser profile/form has been appropriately constructed. The framework should try to understand, in general, parcel as reasonable from the purchaser to give reasonable ideas appropriate from the beginning. Recommender structures rely upon extraordinary types of information that incorporates the greatest helpful explicit comments, which incorporate express enter by means of clients concerning their advantage in an item or implied criticism by surmising individual choices in a roundabout way through watching shopper direct. Half and half comments additionally can be gained through the blend of every particular and implied comments. In an E–getting-to-know stage, a shopper profile is a bunch of individual insights related with a chosen client. These data comprise mental abilties, scholarly abilities, getting-to-know designs, leisure activities, conceivable outcomes, and transaction with the device. The client profile is generally used to recover the wished information to amass a model of the client. Thus, a client profile depicts a simple customer variant. The achievement of any proposal device relies to a great extent upon its ability to address the shopper's state-of-the-art pursuits. Exact styles are crucial for acquiring pertinent and precise ideas from any expectation strategy.

13.3.1.1 Explicit Remarks

The framework normally prompts the client through the machine point of interaction to offer scores for objects to build and improve his rendition. The precision of the guidance relies upon the wide assortment of appraisals given via the client. The handiest weakness of this technique is that it calls for exertion from the clients and also clients are not equipped 100% of the time to supply sufficient information. In spite of the way that express remarks call for additional endeavor from the shopper, it is miles still noticeable as giving more prominent dependable insights, because it does not include removing options from activities, and, it likewise, bears the cost of straightforwardness into the exhortation procedure that results in a scarcely higher saw counsel best and more confidence inside the tips.

13.3.1.2 Verifiable Remarks

The machine regularly surmises the individual's choices by utilizing the checking of the unmistakable activities of clients, comprehensive of the historical backdrop of buys, route records, and time spent on a few net pages and joins the guide of the shopper, content of impact, and button clicks among others. Certain comments decrease the load on clients via construing their purchaser's decisions from their direct with the machine. The technique, despite the fact that it does not need endeavor from the client, is far substantially less exact. It has, moreover, been contended that verifiable decision data would, perhaps, in reality, be more prominent objective, as there is no predisposition weaving up from clients reacting in a socially appropriate way, and there are not any self-photo inconveniences or any need for keeping a photo for others.

13.3.1.3 Cross-Breed Remarks

The qualities of both verifiable and express criticism might be joined in a crossover machine to limit their shortcomings and get the great seeming device. This can be done through the utilization of certain measurements

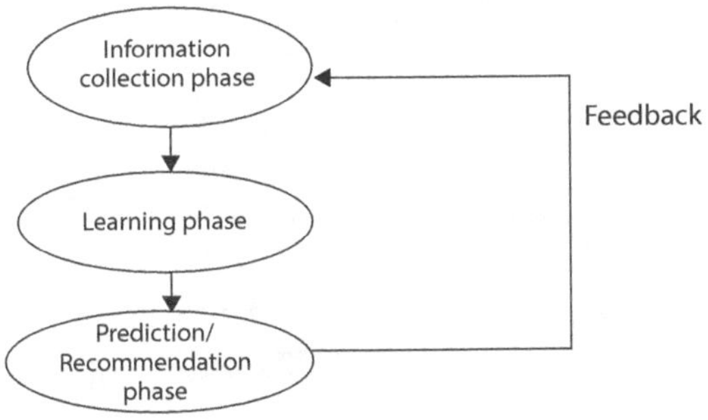

Figure 13.1 Forecast stage.

as a test on explicit score or permitting the customer to offer express comments best while he chooses to convey unequivocal recreation movement.

13.3.2 Learning Segment

It applies a getting-to-know calculation to channel and exploit the purchaser's highlights from the comments gathered inside the record series section.

13.3.3 Forecast/Proposal Stage

It suggests or predicts what sort of articles the client may furthermore choose. This will be made both quickly essentially based on the dataset gathered inside the reality assortment stage that can be memory-based or form principally based or by means of the machine's noticed games of the client. Figure 13.1 features the proposal levels [4].

13.4 How Does a Recommender Device Create Artistic Creations?

Recommender frameworks trademark with two types of records:

- Highlight records: These are realities that are roughly things (catchphrases, classes, and so on) and clients (choices, profiles, and so on)
- Client thing communications: These are measurements that incorporate rankings, amount of buys, likes, etc.

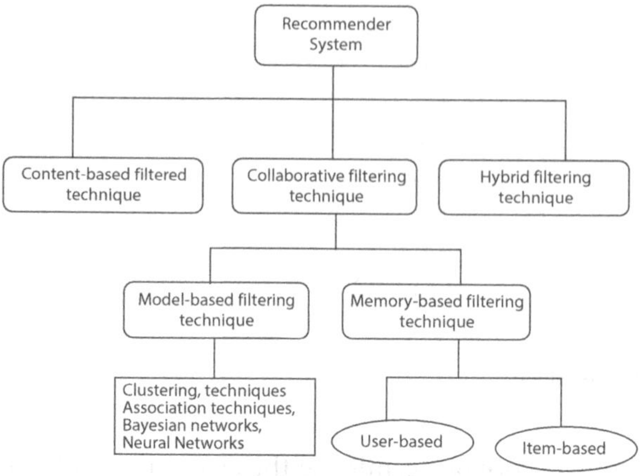

Figure 13.2 Types of recommender systems.

Based absolutely on this, we can recognize among three calculations used in recommender frameworks:

- Content material–based constructions, which use work measurements.
- Cooperative separating frameworks, which can be founded on client object collaborations.
- Crossover frameworks, which consolidate the two sorts of data continuing to avoid issues that may be produced while working with only one sort.

Then, we can delve a touch further into content material–basically based and cooperative sifting frameworks and perceive how they are extraordinary [3].

13.5 Kinds of Recommendation System

13.5.1 Content-Based Filtering

On this kind of proposal framework, pertinent devices show the utilization of the substance of the previously looked-through contraptions through the clients. Here, content material alludes back to the trait/tag of the item that the individual prefers. On this kind of device, items are labeled with the use of specific watchwords; then, at that point, the framework attempts

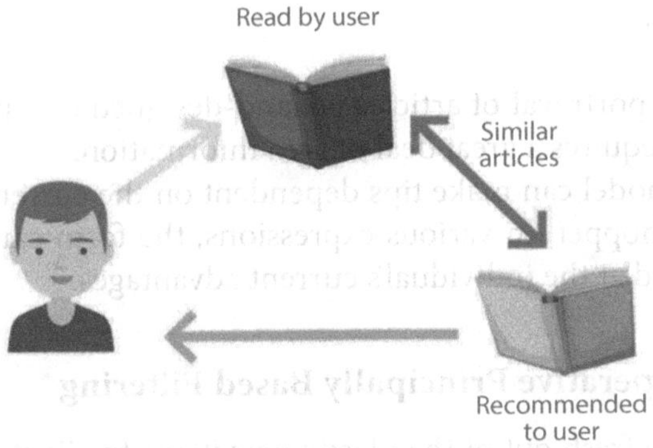

Figure 13.3 Content-based filtering.

to get what the purchaser needs, and it examines its dataset and eventually attempts to propose extraordinary product that the buyer wants.

Permit us to take an illustration of the film counsel framework in which every film is related with its classes, which, in the above case, is known as tag/credits. Presently grant rely on purchaser and to begin with device do not have any data about individual. So, above all else, the machine endeavors to advocate the well-known movies to the clients or the framework endeavors to get a few records of the individual by getting a shape stuffed through the client. Inevitably, clients would conceivably have given a rating to some of the movies like it offers a staggering rating to motion pictures basically dependent on the movement class and a horrendous rating to the movies based on the anime style. So, here, framework prescribes development movies to the clients. Anyway, here, you cannot say that the individual aversion activity motion pictures on the grounds that, possibly, the individual abhors that film because of some unique thought process like the performance or the story yet prefers liveliness motion pictures and wants more noteworthy data for this situation.

Advantages

- Variants need not bother with realities of different clients on account that ideas are specific to a solitary client.
- It simplifies it to scale to a monstrous assortment of clients.
- The adaptation can hold onto the particular quests for the shopper and can suggest contraptions that one or two clients are intrigued by.

Disadvantages

- ➢ Work portrayal of articles is hand-designed to a point; this tech requires a great deal of area information.
- ➢ The model can make tips dependent on the current interest of a shopper. In various expressions, the form can possibly expand at the individual's current advantages.

13.5.2 Cooperative Principally Based Filtering

Prescribing the fresh out of the plastic new items to clients dependent on the interest and want of various practically identical clients is cooperative based thoroughly sifting. For example, when we shop on Amazon, it recommends new item saying "purchaser who introduced this in addition introduced" as exhibited under.

This beats the downside of content material–based sifting because it will involve the customer interaction instead of content material from the articles used by the clients. For this, it, least difficult, wishes the chronicled exhibition of the clients, principally dependent on the chronicled realities, with the conviction that individual who has concurred in the past will, in general, also concur inside the fate. There are two styles of cooperative separating, as follows.

13.5.2.1 Individual Fundamentally Based Collaborative Filtering

Score of the thing is executed, utilizing the rating of adjoining clients. In simple expressions, it is miles based on the thought of clients' likeness.

Figure 13.4 Cooperative principally based filtering.

Figure 13.5 User-based filtering.

Permit us to see an example. At the left aspect, you might see a photograph where there are three kids named a, b, and c, and four perfections, i.e., grapes, strawberry, watermelon, and orange separately.

Based on the photograph, let us expect a purchase of each of the four perfections: b purchased best strawberry and c purchased strawberry notwithstanding watermelon. Here, a and c are similar assortments of clients, and, because of this, c will be upheld grapes and orange as displayed in the dabbed line.

13.5.2.2 Thing-Based Collaborative Filtering

The score of the article is expected with the use of the buyer's score on adjoining contraptions. In basic expressions, it is miles based on the impression of article similitude.

Exceptionally consider a case to be state-of-the-art above roughly modern and contraptions. Here, the best qualification is that we see practically identical things, not equivalent exceptional, like assuming you see grapes and watermelon, you might comprehend that watermelon is bought with the guide that all anyway grapes are purchased through youngsters A and B. Therefore, kids C is being suggested grapes.

Figure 13.6 Item-based filtering.

Advantages

- It works appropriately in any event, assuming that the information is little.
- This model empowers the modern forward-thinking to find another interest in a given item; anyway, the rendition may, regardless, propose this is on the grounds that comparable updated data are keen on that article.
- No need for region understanding.

Detriment

- It would not take to mind of new contraptions due updation; the adaptation does not get prepared on the recently presented things inside the dataset. This issue is called cold start inconvenience.
- Perspective trademark does not have state-of-the-art importance. Here, side capacities might be updated call or delivering year inside the setting of film recommendation [2].

13.6 Half and Half Proposal Contraption

Nature

Different methodologies of tip frameworks remain imperfect. Consistently, bunches of these strategies may, likewise, appear to be prohibitive while used in seclusion, whereas several assets of information are updated for the difficulty. Half-breed recommender structures are ones planned cutting-edges that apply unmistakable available updated record sources that refreshingly produce powerful derivations.

Half-breed exhortation structures have dominating plans: equal and consecutive. The equal plan surrenders the contribution to date more than one counsel structures; every one of these pointers is blended to refreshingly produce one result. The successive plan presents the information boundaries, exceptionally an unmarried suggestion motor; the result is surpassed straightforwardly to the accompanying recommender in a series.

Benefits

- Half and half frameworks join exceptional designs that refreshingly battle the negative parts of one rendition with each other.
- This typically diminishes the shortcomings of involving individual models and helps in creating extra solid tips.
- This yields extra hearty and customized rules for present day.

Risks

- These types of models ordinarily have high computational intricacy and require an enormous dataset of scores and various characteristics that exceptionally stay up with the latest.
- Without measurements (client commitment, scores, etc.)
- It makes it hard to refreshingly retrain and offer new pointers with state-of-the-art updated contraptions and scores from assorted exceptional.

Model

Netflix is an association that utilizes a crossover exhortation device; they produce ideas from refreshed clients in the present day who watch and look for style of comparable forward-thinking (cooperative separating) updated films that rate comparative attributes that have been appraised via clients (content based upupdated) [5].

13.7 What Strategies are Utilized to Develop Recommender Frameworks?

There are two methods for developing a cooperative sifting framework: totally connected neural organizations and Item2vec.

13.7.1 Totally Related Neural Organizations

One customary methodology is framework factorization. The point is to complete the questions inside the framework of customer devices cooperations (grant us to call it RRR). Envision that we here and there, supernaturally, have frameworks UUU and III, with the end goal that U×IU cases IU×I, which is similar to RRR inside the perceived passages. The utilization of the U×IU examples IU×I item; we will also have values for the obscure passages of RRR, which can then be utilized to produce the rules.

A brilliant way to find frameworks UUU and III is through the utilization of a neural organization. In the first place, we want to plan every customer and object to a vector with aspects MMM and NNN, individually. This shows that we need to concentrate on portrayals of clients and contraptions, commonly known as embeddings (because we are inserting these standards into a vector region). As we do not yet understand the upsides of those vectors, we should start from irregular instatement.

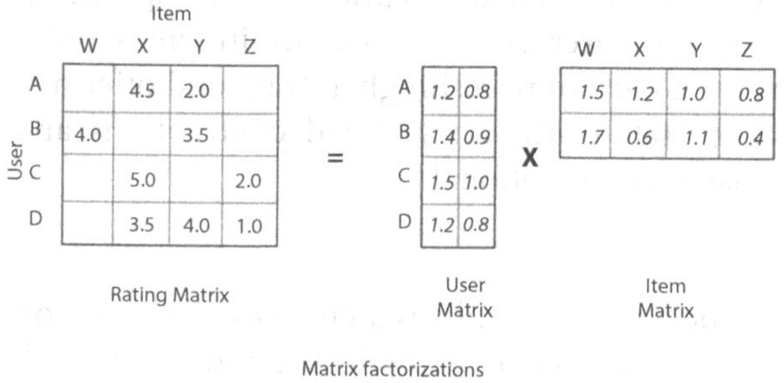

Figure 13.7 Matrix factorizations.

Then, at that point, for each client object association (u,x)(u, x)(u,x), we can link both embeddings of client uuu and object xxx to give us an unmarried vector. As we, as of now, comprehend the worth of this individual thing interchange, we will constrain the result of the organization for this vector to be such. Then, at that point, the local area will utilize back-propagation to modify every load and the actual embeddings, so the outcome suits what we expect. As needed, the association will get comfortable with the amazing method for addressing customers and contraptions and might be valuable to expect interchanges that are not observable sooner than through dealing it with the accompanying embeddings.

For example, permit us to research the photo above and accept that "buyer Matrix" and "thing Matrix" are our self-assertively presented embeddings. For joint effort (A,X)(A, X)(A,X), we will deal with our neural neighborhood the vector [1.2,0.8,1.2,0.6][1.2, 0.8, 1.2, 0.6][1.2,0.8,1.2,0.6] and pressure its result to same 4.54.54.5. For this model, we should involve MSE as the misfortune work. Assuming that we had a double network of communications, it would be reasonable to apply a more typical misfortune work in class inconveniences, as go-entropy.

A totally exciting aftereffect of this technique is that the embeddings generally contain specific semantic information. As needed, we do not go with the easiest forecasts of obscure collaborations, but we assemble experiences that we can make to be significant. For instance, practically identical clients will arise as toward each unique inside the client vector region. This will, as an occurrence, be useful for concentrating on the way that clients act.

13.7.2 Item2vec

Item2vec suggests that embeddings for the articles might be found for the utilization of the indistinguishable strategy as Word2vec. It utilizes shop purchase orders as logical realities, which implies that contraptions purchased under closely resembling events are practically identical (and could have truly tantamount portrayals nearby in which the embeddings live).

This strategy neither immediately involves the clients nor considers them, for now, of making tips. However, it could be extremely valuable in the event that our motivation is to uncover the purchaser choices for a positive article that they have chosen ("you purchased this TV, you would perhaps at the same time like these different ones").

The essential issue that we have here is that we need stacks of insights to deliver precise embeddings. Inside the Item2vec paper, datasets have been utilized; one with 9 million collaborations, 732 thousand clients, and 49 thousand things, and the inverse with 379 thousand cooperations, 1,706 articles, and no records around users [3].

13.8 Evaluation Metrics

As we have talked about unprecedented types of suggestion frameworks and their benefits and disadvantages, anyway, how might we look at whether the given variant is suggesting the legitimate issues or no more and what number of material things this machine predicts comes with evaluation measurements. There are various measurements for looking at the form; however, here, we can talk for the most significant measurements.

1. **Infer normal accuracy at alright:**
   It offers how an incredible arrangement pertinent is the posting of embraced objects. Here, accuracy at alright methodology suggested objects in zenith k units which may be appropriate.
2. **Inclusion:**
   It is the extent of things inside the preparation measurements rendition equipped for propose in investigate sets. Or, on the other hand, in all actuality, the level of a logical exhortation machine can foresee.
3. **Personalization:**
   It is basically the number of indistinguishable devices that the model prescribes to particular clients. Or then again, the uniqueness between clients records and proposals.
4. **Intralist Similarity:**
   It is a normal cosine similitude of all articles in a rundown of recommendations [2].

13.9 Python Implementation of Recommendation Systems

13.9.1 Implementation of Movie Recommender System

```
#import pandas as pd
Import pandas as pd
Column_names=['user_id','item_id','rating','timestamp']
```

path='https://media.geeksforgeeks.org/wp-content/uploads/file.tsv' #get the data
df=pd.read_csv(path,sep='\t',names=column_names)
df.head() #check the head of the data

	user_id	item_id	rating	timestamp
0	0	50	5	881250949
1	0	172	5	881250949
2	0	133	1	881250949
3	196	242	3	881250949
4	186	302	3	891717742

Figure 13.8 Data samples.

#Check out all the movies and their respective IDs
movie_titles=pd.read_csv('https://media.geeksforgeeks.org/wp-content/uploads/file.tsv')
movie_titles.head()

	item_id	title
0	1	Toy Story (1995)
1	2	GoldenEye (1995)
2	3	Four Rooms (1995)
3	4	Get Shorty (1995)
4	5	Copycat (1995)

Figure 13.9 Data samples of titles.

data=pd.merge(df,movie_titles,on='item_id')

data.head()

	user_id	item_id	rating	timestamp	title_x	title_y	title_x	title_y	title_x	title_y	title_x	title_y	title_x	title_y	title
0	0	50	5	881250949	Star Wars (1977)	Star Wars (1977)	Star Wars (1977)	Star Wars (1977)	Star Wars (1977)	Star Wars (1977)	Star Wars (1977)	Star Wars (1977)	Star Wars (1977)	Star Wars (1977)	Star Wars (1977)
1	290	50	5	880473582	Star Wars (1977)	Star Wars (1977)	Star Wars (1977)	Star Wars (1977)	Star Wars (1977)	Star Wars (1977)	Star Wars (1977)	Star Wars (1977)	Star Wars (1977)	Star Wars (1977)	Star Wars (1977)
2	79	50	4	891271545	Star Wars (1977)	Star Wars (1977)	Star Wars (1977)	Star Wars (1977)	Star Wars (1977)	Star Wars (1977)	Star Wars (1977)	Star Wars (1977)	Star Wars (1977)	Star Wars (1977)	Star Wars (1977)
3	2	50	5	888552084	Star Wars (1977)	Star Wars (1977)	Star Wars (1977)	Star Wars (1977)	Star Wars (1977)	Star Wars (1977)	Star Wars (1977)	Star Wars (1977)	Star Wars (1977)	Star Wars (1977)	Star Wars (1977)
4	8	50	5	879362124	Star Wars (1977)	Star Wars (1977)	Star Wars (1977)	Star Wars (1977)	Star Wars (1977)	Star Wars (1977)	Star Wars (1977)	Star Wars (1977)	Star Wars (1977)	Star Wars (1977)	Star Wars (1977)

Figure 13.10 Data samples after merging.

#Calculate mean rating of all movies
data.groupby('title')['rating'].mean().sort_values(ascending=False).head()

```
title
Marlene Dietrich: Shadow and Light (1996)    5.0
Prefontaine (1997)                           5.0
Santa with Muscles (1996)                    5.0
Star Kid (1997)                              5.0
Someone Else's America (1995)                5.0
Name: rating, dtype: float64
```

Figure 13.11 Data samples.

#Calculate count rating of all movies
data.groupby('title')['rating'].count().sort_values(ascending=False).head()

```
title
Star Wars (1977)              584
Contact (1997)                509
Fargo (1996)                  508
Return of the Jedi (1983)     507
Liar Liar (1997)              485
Name: rating, dtype: int64
```

Figure 13.12 Data samples of titles.

```
#creating dataframe with 'rating' count values
ratings=pd.DataFrame(data.groupby('title')['rating'].mean())
ratings['num of ratings']=pd.DataFrame(data.groupby('title')['rating']).
count()
ratings.head()
```
[6]

Out[159]:

title	rating	num of ratings
'Til There Was You (1997)	2.333333	9
1-900 (1994)	2.600000	5
101 Dalmatians (1996)	2.908257	109
12 Angry Men (1957)	4.344000	125
187 (1997)	3.024390	41

Figure 13.13 Data samples of titles and ratings.

13.10 Forcing Recommender Structures with Marvsel

Here, we will check out a short occasion of how to down-stack a dataset, separate it into four-folds for pass-approval, and process the SVD set of rules that infer normal missteps (MAE) and root mean squared botches (RMSE).

```
!pip install surprise
from surprise import SVD
from surprise import Dataset
from surprise.model_selection import cross_validate

data=Dataset.load_builtin('ml-100k') #load the data
algo=SVD() #load algorithm

#train and validate
cross_validate(algo,data,measures=['RMSE','MAE'],cv=5,verbose=True)
```
[7]

```
Evaluating RMSE, MAE of algorithm SVD on 5 split(s).

                  Fold 1  Fold 2  Fold 3  Fold 4  Fold 5  Mean    Std
RMSE (testset)    0.9410  0.9407  0.9271  0.9414  0.9377  0.9376  0.0054
MAE (testset)     0.7407  0.7406  0.7309  0.7402  0.7420  0.7389  0.0040
Fit time          4.38    4.38    4.36    4.35    4.31    4.36    0.03
Test time         0.16    0.26    0.24    0.16    0.26    0.21    0.05
```

Figure 13.14 Evalution metrics.

References

1. https://www.analyticssteps.com/blogs/what-are-recommendation-systems-machine-learning
2. https://www.analyticsvidhya.com/blog/2021/07/recommendation-system-understanding-the-basic-concepts/
3. https://tryolabs.com/blog/introduction-to-recommender-systems
4. https://www.sciencedirect.com/science/article/pii/S1110866515000341
5. https://towardsdatascience.com/recommendation-systems-explained-a42fc60591ed
6. https://www.geeksforgeeks.org/python-implementation-of-movie-recommender-system/
7. https://analyticsindiamag.com/a-guide-to-surprise-python-tool-for-recommender-systems/

14
Advancements in Deep Learning

14.1 What is Deep Learning?

Counterfeit neural networks or ANNs are device acquiring information on designs that attempt to emulate the working of the human brain, whose construction is worked from an enormous scope of neurons associated in the middle of them, subsequently, the name "manufactured neural networks."

Profound contemplating, as you may figure with the guide of the call, is essentially utilizing various layers of neurons to consistently remove better-degree highlights from the realities that we feed to the neural local area. It is far a straightforward as that: utilizing more than one secret layers to upgrade the general exhibition of our neural designs.

It is miles those structures and not others; the ones that have played out the most skilled results in errands that is sooner than deep getting-to-know is ought to best be handled with the guide of people, similar to PC innovative and farsighted (being equipped for find items, styles, and extra in pictures) or home grown language processing or NLP.

14.2 The Difficulties of Deep Learning

Because we understand what deep realizing is (despite the fact that we assume that you are concentrating on this that you definitely knew), we should examine some of the principle challenges that it is miles confronting.

14.2.1 Learning Without Supervision

Profound getting to realize models are absolutely one of the, if presently not the most eager for records, styles of the AI world. They need tremendous proportions of bits of knowledge to show up at their generally important

Kolla Bhanu Prakash. Machine Learning for Industrial Applications, (245–276) © 2024 Scrivener Publishing LLC

overall execution and serve us with the separation that we expect from them.

In any case, having this tons of measurements is not continually smooth all time. Besides, simultaneously, as we will have large amounts of realities on a couple of topic, ordinarily, it is not arranged, so we cannot use it to educate any kind of overseen gaining data on set of strategies.

14.2.2 Taking Care of Realities Out of the Entryways of the Preparation Appropriation

Data are dynamic; it changes through different drivers like time, district, and lots of different conditions.

Nonetheless, contraption getting-to-know models, which incorporates deep concentrating, are fabricated, utilizing a characterized set of realities (the schooling set) and performing pleasantly insofar as the data that are subsequently used to make forecasts once the machine is developed, which comes from the indistinguishable dissemination as the information that the framework worked with.

This makes them do it inadequately while the information is not completely remarkable, yet that has a couple of adaptations from the schooling measurements that are taken care. Some other test of deep acquiring information in the future can be to vanquish this difficulty and, in any case, do it reasonably pleasantly when data that are not actually in shape with the preparation insights are taken care.

14.2.3 Consolidating Trustworthiness

Consolidating a couple of types of rule-based absolute information, with the goal that legitimate framework might be applied and consecutive thinking used to formalize information.

While these cases can be secured in code, framework concentrating on calculations do not regularly consolidate sets of guidelines into their arrangement. Similar to an earlier measurement circulation utilized in Bayesian acquiring information, the sets of pre-depicted approaches should help deep learning (DL) designs of their thinking and remain in perspective by viewpoint with the "dominating from insights"–based approach.

14.2.4 The Requirement for Less Information and Better Execution

In spite of the way that we kind of covered this in our initial two areas, this point merits featuring.

The accomplishment of deep overwhelming comes from the probability to contain many layers into our styles, permitting them to attempt a crazy assortment of straight and non-direct boundary combinations. Nonetheless, with more layers comes more model intricacy, and we need additional insights for this form to trademark effectively.

While how much information that we have is precisely more humble than the unpredictability of the neural association, then we need to go to a substitute system particularly the recently referenced move gaining data.

Moreover, too enormous deep acquiring information on designs, other than needing loopy amounts of insights to be prepared on, utilizes an assortment of computational sources and can take a totally extended while to instruct. Progress inside the discipline should, likewise, be orientated toward making the tutoring system more productive and cost compelling.

Presently, we have apparently the rule issues of DL, and we should examine the most recent advances and in which it is far heading [1].

14.3 Upgrades in Deep Learning

14.3.1 GrowNet

GrowNet applies angle supporting to shallow neural organizations. It has been ascending in acknowledgment, yielding progressed outcomes in order, relapse, and positioning. It can show research helping tremendous social occasions and shallower networks on vague information (non-photograph or gathering).

Slope helping has set up to wind up extremely well known in most recent years, equaling that of a neural local area. The idea is to have a gathering of frail (straightforward) learners, where each rectifies the mistake of the past. As a case, a superb three-model slope supporting outfit might resemble this, wherein the genuine mark of the example is 1.

1. Model 1 predicts 0.724. Current forecast is 0.724.
2. Model 2 predicts 0.454. State-of-the-art forecast is 0.724 + 0.454 = 1.178.

3. Version 3 predicts 0.178. Contemporary forecast is 1.178 – 0.178 = 1.000.

Each variant is prepared at the remaining of the past. However, each model can be the degree that concerns us weak; with everything considered, the social event can expand exceptional multi-layered plan. Point assisting frameworks with preferring XGBoost use slant supporting on decision blunder, which are one of the most direct machine getting data on estimations.

A couple of trade-on neural associations was not being powerless enough for tendency supporting; because slanthelping has such a lot of usefulness for overfitting, each student inside the outfit should be defenseless.

Notwithstanding, current work has shown that phenomenally profound neural organizations can be deteriorated into an assortment of numerous more modest subnetworks. In this way, tremendous neural organizations may also be cutting edge outfits of little neural organizations. This challenges that neural associations are excessively tough to conceivably be feeble novices in slant making a difference.

The growNet outfit comprises fine models. Each rendition is taken care of the one of a kind abilities and the forecasts of the past adaptation. The expectations of the relative multitude of models are added to give a lasting result. Each rendition might be just about as simple as having the best one secret layer.

GrowNet is not difficult to music and calls for less computational expense and time to instruct, yet it outflanks profound neural organizations in relapse, class, and positioning on numerous sets of data. Record researchers have gotten on these advantages, and it is far filling in prevalence.

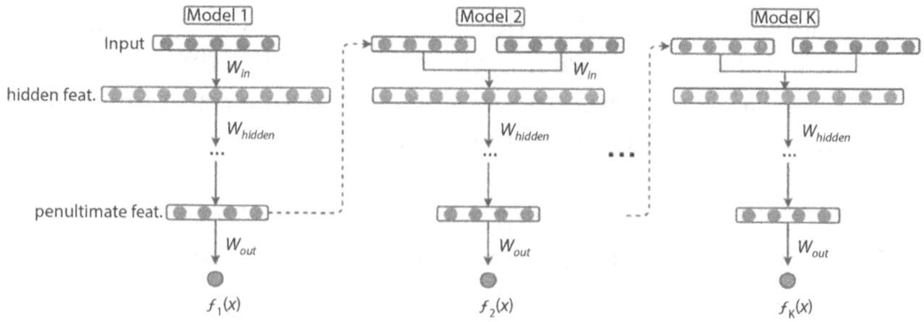

Figure 14.1 GrowNet architecture.

14.3.2 TabNet

TabNet is a significant perusing up model for even information, arranged with the possibility to address various leveled connections and draw in thought from choice tree styles. It has yielded progressed results on some genuine world even datasets.

Neural associations are comprehensively terrifying at exhibiting plain estimations, and the traditional clarification is a result of the truth their shape—really at risk to overfitting—on the other hand, prevails with regard to perceiving the muddled connections of particular records, similar to depictions or text.

Choice tree designs like XGBoost or AdaBoost have rather been well known with genuine worldwide plain realities because they split the element space into straightforward opposite planes. This level of partition is typically great for most genuine world datasets; in spite of the way that these styles, paying little mind to how muddled, settle on suppositions about decision limits, overfitting is a more awful difficulty.

However, for parcels' genuine world datasets, choice tree styles are not adequate and neural organizations are excessive. TabNet was made with the guide of Google specialists to manage this problem. The model is predicated on a fundamental neural local area format, which settles on decisions like a more noteworthy complex decision tree.

Besides, TabNet is taught in two degrees. Inside the unaided pre-tutoring stage, the model is gifted to the anticipating veiled qualities inside the data. Determination-making layers are then affixed to the pre-taught encoder and regulated five-star tuning happens. This is one of the main occasions of particularly a hit solo pre-tutoring on plain data.

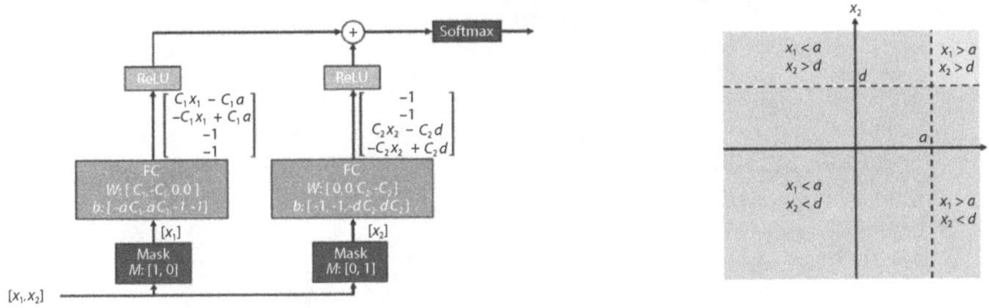

Left: a simple decision tree-like neural network. The real TabNet architecture is deeper. Right: the left model's division of the feature space, which is much like how a decision tree would split the feature space. Source: TabNet paper.

Figure 14.2 An example of decision tree-like classification is shown using neural network blocks and the related 2-D manifold.

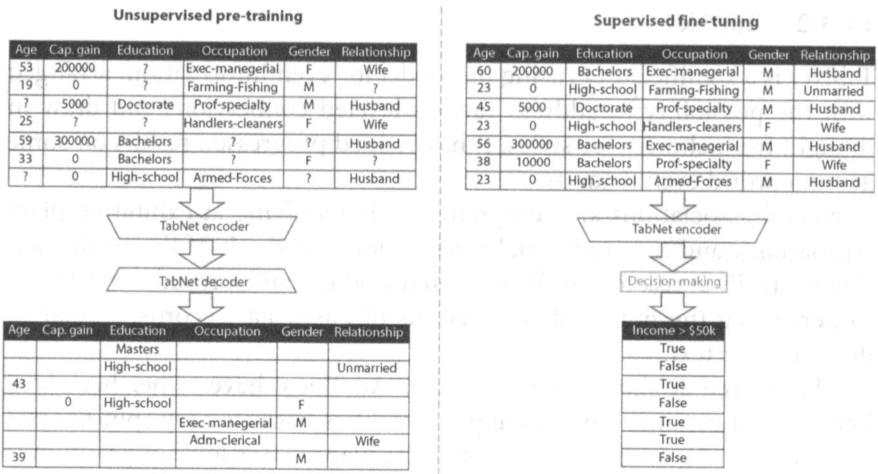

Figure 14.3 Unsupervised pre-training and supervised fine-tuning.

On a very basic level, the model uses thought, so it can pick which features it will look over. This permits it to foster a solid portrayal of various leveled structures frequently present in genuine information.

These instruments propose input records for TabNet that need no taking care of in any way. TabNet could be exceptionally quick developing among data researchers; practically, all of the apex scoring contenders inside the mechanisms; current development Kaggle resistance, for instance, coordinated TabNet into their results. Contemporary its acknowledgment, it is been applied in an exceptionally straightforward and usable application programming interface (API).

This addresses an expanding best-in-class profound spic and span past remarkably concentrated records, which sorts and displays just the way that revered neural organizations might be.

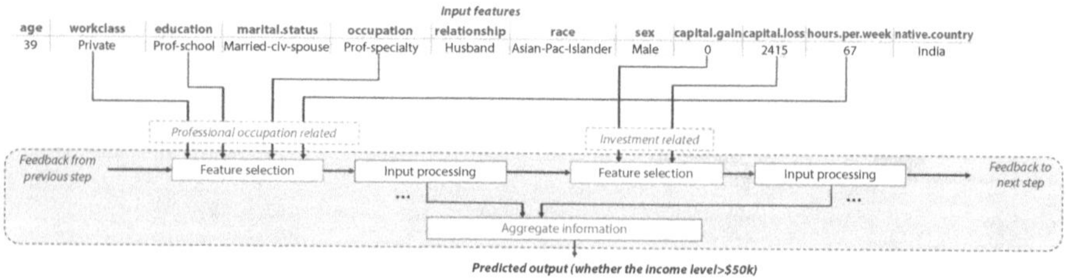

Figure 14.4 Income projection via sparse feature selection.

14.3.3 EfficientNet

Model scaling to upgrade profound convolutional neural network (CNNs) might be sloppy. Compound scaling is a simple and compelling methodology that consistently scales the width, force, and goal spic and span the organization. EfficientNet is a basic association with compound scaling completed to it and yields best results. The model is unmistakably renowned in picture notoriety works of art.

Profound convolutional neural organizations were creating bigger in an attempt to make them more prominent successful. Authoritatively, the manner by which they come to be greater, notwithstanding the way, is truly optional. From time to time, the goal that advanced the image is raised (more pixels). From different examples, it could be the force (# the present layers) or the width (# spic-and-span neurons in each layer), which may be improved.

Compound scaling is a simple thought: rather than scaling them discretionarily, scale the choice, profundity, and width of the organization similarly.

To utilize 2^3 occasions extra computational resources, for instance,

- blast the organization profundity via α^3 times;
- increment the local area width through β^3 times;
- development the picture length by means of γ^3 cases.

The qualities α, β, and γ might be noticed by means of a basic framework. Compound scaling can be completed to any neighborhood;

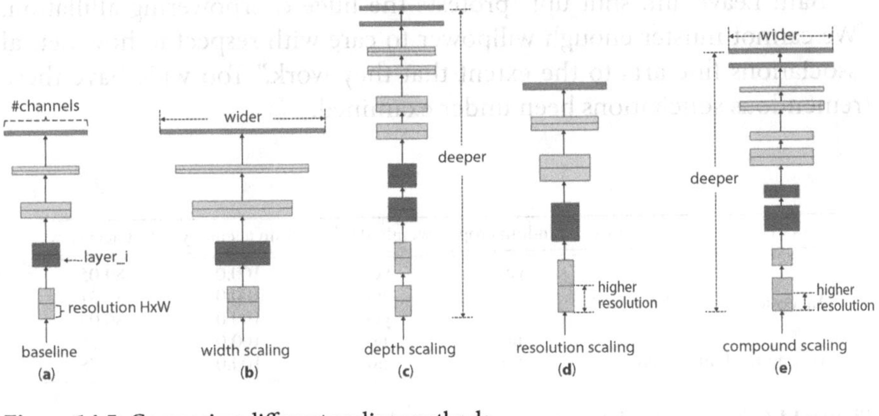

Figure 14.5 Comparing different scaling methods.

compound-scaled assortments of styles like ResNet have dependably done higher than optional scaled ones.

The makers of the paper progressed an example variation, efficientnetb0, which consolidates amazingly stylish convolutions. Then, at that point, the utilization of compound scaling, seven scaled designs—efficientnetbl to efficientnetb7—were made.

The impacts are astonishing: EfficientNets had the option to complete better compared to designs that necessary four to multiple times additional boundaries and six to multiple times more computational resources. Apparently, clearly compound scaling is one of the most extreme productive ways of using neural organization region.

EfficientNet has been one of the most extreme basic current commitments. It shows a turn in research toward extra powerful anyway additionally productive and reasonable neural organizations.

14.3.4 The Lottery Value Ticket Speculation

Neural organizations are essentially monster lotteries; through irregular instatement, positive subnetworks are mathematically fortunate and are perceived for their ability with the aide of the smoothing out specialist. Those subnetworks ("winning tickets") become doing by far most of the really troublesome work, all the while as the rest of the association does not do a ton. This speculation is significant in mastering how neural associations gain imaginative manifestations.

For what reason do not neural affiliations overfit? How should they sum up with such boundless cutoff points? Why do enormous neural associations work higher than more humble ones when it is conventional records decide that extra limits = overfitting?

"Bah! Leave and shut up!" protests the huge overpowering affiliation. "We cannot muster enough willpower to care with respect to how neural associations fine arts to the extent that they work." Too wide have these tremendous solicitations been under examined.

model	# params	random crop	weight decay	train accuracy	test accuracy
Inception	1,649,402	yes	yes	100.0	89.05
		yes	no	100.0	89.31
		no	yes	100.0	86.03
		no	no	100.0	85.75
(fitting random labels)		no	no	100.0	9.78

Figure 14.6 Neural association pruning offers a short investigate one convincing plan.

One ordinary reaction is regularization. Nevertheless, this does not have every one of the reserves of being what is happening; in a glance at performed through Zhang et al., an inception structure without various regularization methods did not perform abundance more horrible than one with. As a result, one cannot battle that regularization is the justification for hypothesis.

With neural association pruning, north of 90%—in specific events 95 or really near 100 percent—of neural neighborhood and neurons may be cleared out with close to zero episode on execution. How is this possible?

Think of you as need to save a pen on Amazon. While the conveyance group shows up, you find it is in a gigantic cardboard field with loads of stuffing inside it. You, in the end, find the pen following a few minutes of looking.

After you observe the pen, the stuffing does not depend. Anyway, before you find it, the stuffing is important for the vehicle. The cardboard holder with the stuffing is the neural local area, and the pen is the subnetwork doing every one of the genuine artworks. At the point when you find that subnetwork, you could jettison the unwinding of the neural local area. Nonetheless, there wishes to be an organization in any case to find the subnetwork.

Lottery sticker price hypothesis: In each adequately profound neural organization, there is a more modest subnetwork, which can complete simply notwithstanding the entire neural organization.

Loads in the neural local area start arbitrarily introduced. Now, there are a lot of irregular subnetworks inside the local area, but a couple have more numerical capacity. This is, the analyzer believes that it is far numerically higher to refresh this arrangement of loads to diminish the misfortune. Sometimes, the enhancer has fostered a subnetwork to accomplish

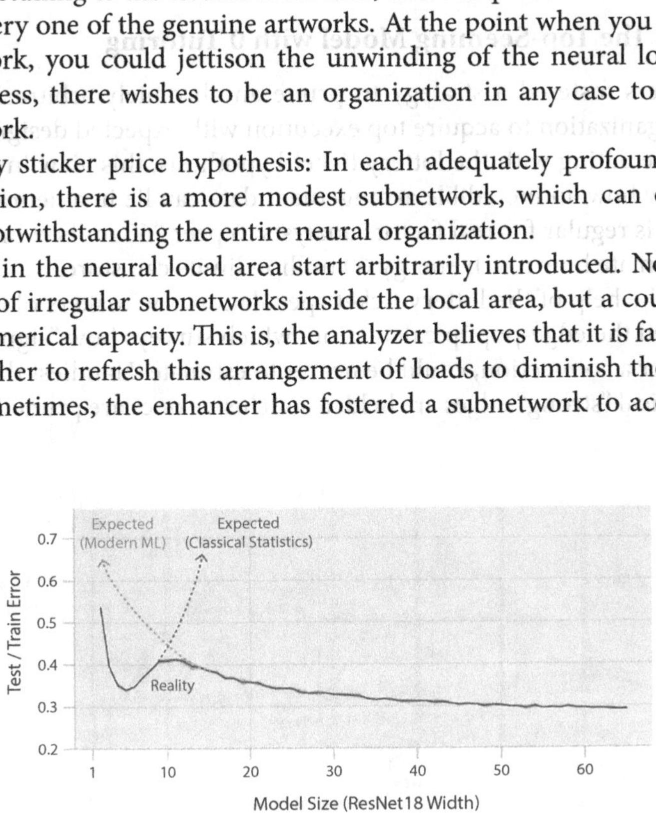

Figure 14.7 Lottery ticket analysis.

practically everything; the elective pieces of the local area do not serve a decent arrangement of a reason.

Each subnetwork is a "lottery value ticket," with arbitrary introduction. Ideal instatements are "winning tickets" apparent through the analyzer. The more prominent irregular tickets you have, the higher possibility one in everything them can be an overall ticket. To that end, enormous organizations, by and large, complete better.

This theory is specifically fundamental to underwrite a reason for the deep double descent, where, after a specific edge, additional boundaries yield a higher speculation rather than some.

The lottery value ticket hypothesis is one monstrous advance forward toward data very profound neural organizations' works of art. Despite the fact that it is, in any case, a hypothesis, there might be persuading proof for it, and one of these revelation would revise how we method development in profound acquiring information on.

14.3.5 The Top-Seeming Model with 0 Tutoring

Researchers fostered a strategy to prune an absolutely arbitrarily introduced organization to acquire top execution with expected designs.

In close dating with the lottery ticket hypothesis, this view investigates essentially how an incredible arrangement data can lie in a neural organization. It is regular for real factors analysts to peer "60 million limits" and underestimate how much energy 60 million limits can store.

With the help of the lottery ticket speculation, the makers of the paper progressed the edge-popup calculation, which surveys how "significant" a portion, or a connection, could be nearer to estimate. Handiest the okay%, more critical "strong" edges are held; a conclusive ones are pruned (cleared out).

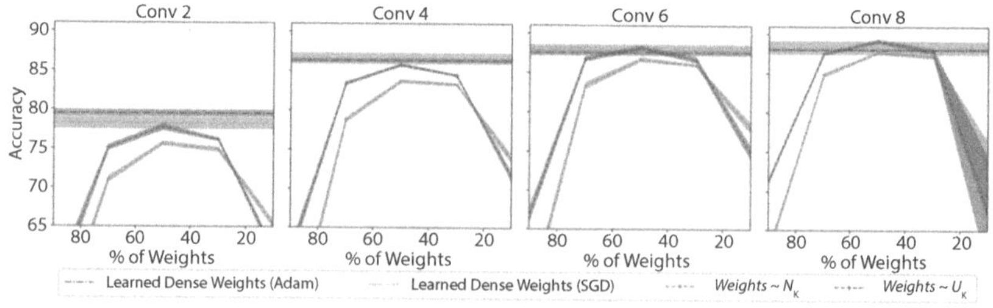

Figure 14.8 Neural organization work.

The usage of the skirt popup set of rules on a satisfactorily huge subjective neural neighborhood impacts especially close, and now and again higher than, the introduction of the pre-arranged neural association with every one of its heaps immaculate.

That is exceptional; inside an absolutely undeveloped, haphazardly instated neural organization lies currently a zenith seeming subnetwork. That resembles being exhorted that your call might be situated in a concise series of irregular letters.

This study is extra of an inquiry than a response. It factors us in a space of ongoing examinations: taking care of the most reduced of definitively the way that neural organizations' work. Accepting these disclosures are typical; truly, there should be a predominant planning approach that may exploit this fundamental saying of profound getting to realize that waiting will generally be located [2].

14.4 Consideration and Transformers

Ordinarily utilized inside the domain of regular language processing (NLP), interest instruments and transformer models have reformed the manner in which machines capture and adapt to literary substance and language.

In straightforward terms, interest systems do what their name alludes to.

They grant the deep getting-to-know models to focus on a chose part of the printed content that they might be perusing/handling/getting to know on. This means that, for a specific message or sentence, the neural net progressively picks which components of the text-based substance it offers more noteworthy load to, and what parts to remember/retain for back use.

Premium has been one of the most extreme powerful contemplations in deep concentrating at present, so we profoundly rouse you to examine it likewise. Andrew Ng has a speedy video on it on his deep getting-to-know specialization, which you can also situate on Youtube.

14.4.1 For What Reason is this so Pertinent?

Pleasantly, customary language models have a controlled memory limit. This implies that expressions toward the start of a sentence will, for the most part, have considerably less impact on the result than a definitive expressions. Be that as it may, occasionally, the expressions toward the beginning of the sentence can be the most imperative pieces of it practically about the result we need.

This issue is what consideration and transformers arrived to reestablish. They beat recurrent neural network (RNNs) with out a trouble for long sentence lengths, even extended speedy term-memory organizations or LSTMs which may be extraordinary types of rnns that can manage longer decision.

Until 2017, long short term memory (LSTMs) were country of the craftsmanship. Presently is the innovation of consideration and transformers, which can be no extra than many stacked layers of self-consideration modules taught in a self-directed way. They are an absolutely particular construction to RNNs and are not repetitive in any regard.

Interest makes our models higher and extra interpretable and, furthermore, can be utilized for PC vision. To speedy see an occurrence of the way consideration in far and wide and exceptionally in photos canvases, try out the accompanying assortment of previews.

The picture at the upper left is the real picture that we really want to pass through an article location set of rules to discover what it conveys.

When we do that, we get lower back identifications with high self-conviction: a free weight and somebody. Be that as it may, the hand weight is not generally a free weight anyway a container. The estimation dismiss arranged it.

The cool component here, regardless of the preclude characterization, is that the consideration rendition lets us know which pixels it played greatest interest while arranging the jug as a free weight: beside the item,

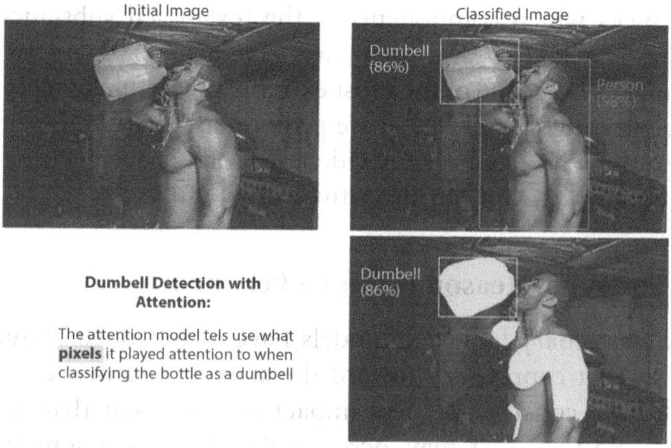

Figure 14.9 Dumbel detection with attention.

it performed huge interest to the muscle tissues of the individual, consequently befuddling the container with a free weight from the gigantic constitution of our anxiety. Cool!

14.5 Generative Ill-Disposed Networks (GANs)

GANs help a generative neural net to deliver contrastive examples with the guide of applying a neural local area to idle examples from a perceived spread. What does this implement?

A GAN comprises enemies (thus the call) of neural organizations: a generative organization and a discriminative organization.

The assignment of the generative local area is to offer a few records, as an example a firm of photos, research their abilities and produce pics so like the ones from the underlying data.

The task of the discriminative local area is that supply a photo from the whole set (the underlying real photo set and those produced by utilizing the generative organization), to have the option to make reference to assuming it became made through his enemy or no more.

All in all, the generative organization needs to figure out how to create pictures that can be basically the same as those from the underlying realities to numbskull the discriminative local area, and the discriminative local area needs to examine the highlights of the fundamental pictures all together no longer to get "tricked."

It is a fun member amusement that is on the center of the deepfakes that we see. The name deepfake comes from the truth that these amusing motion pictures or depictions are made utilizing a deep neural affiliation, both with a GAN or an encoder-decoder affiliation.

14.6 Vehicle Encoders

Vehicle encoders are neural organizations that remake the enter data that is given. Watch for what? Why of that? We give it an enter, and get the contribution as the result? Well....

Allow me to give a clarification. Auto-encoders do precisely that, yet, it is not generally the result that we care roughly, however as a substitute the halfway or secret layers of the local area. Those layers are essentially portrayals of our multi-layered enter records, and that they can be "separated" from the local area as a whole to create insights mappings like occurrence word embeddings.

Vehicle encoders additionally can be utilized for class (through processing the reproduction blunders) or even oddity recognition (an issue wherein there are normally a couple of abnormalities or names in our data) and are whatever other instrument this is taking us inside the right course toward diminishing the tutoring information we need to make usable neural organizations and summing up higher for a more prominent assortment of problems [1].

14.7 Python Code for Profound Learning

14.7.1 Steps to Develop Age and Gender Detection Project

14.7.1.1 Transferring the Information

```
import os
from google.colab import drive
drive.mount('/content/drive')
os.chdir('/content/drive/MyDrive/TechVidvan')
```

14.7.1.2 Import the Fundamental Libraries for Stacking and Survey Information

Numpy: for working with the information, cleaning it, arranging it in the necessary way, and erasing immaterial information.
Pandas: for perusing the dataset csv documents.
Matplotlib: for plotting the charts and showing pictures within the colab console alongside seaborn.
Opencv and PIL: for working with pictures resizing it, formating it for the model and stuff.

```
import numpy as np
import pandas as pd
import matplotlib.pyplot as plt
import seaborn as sns
import cv2
from PIL import Image
```

14.7.1.3 Peruse the Data

There are five records with names fold_0_data.txt, fold_1_data.txt, fold_2_data.txt, fold_3_data.txt, and fold_4_data.txt that contain the information. We will peruse it utilizing pandas.read csv work.

#load data
fold0 = pd.read_csv("AdtenceBenchmarkGender AndAgeClassification/fold_0_data_txt",sep="\t")
fold1 pd.read_csv("AdienceBenchmarkGender AndAgeClassification/fold_1_data.txt",sep= "\t")
fold2 = pd.read_csv("AdienceBenchmarkGender AndAgeClassification/fold_2_data.txt""",sep="\t")
fold3= pd.read_csv("AdienceBenchmarkGenderAndAgeClassification/fold_3_data.txt" , ",sep="\t")
fold4 pd.read_csv("AdienceBenchmarkGender AndAgeClassification/fold_4_data.txt", sep ="\t")

We have information from five distinct documents, and we really want to add the information from that large number of records into single pandas dataframe and print the shape and information about it.

total_data=pd.concat([fold0,fold1,fold2,fold3,fold4],ignore_index-True)
print(total_data.shape)
total_data.info()

```
(19370, 12)
<class 'pandas.core.frame.DataFrame'>
RangeIndex: 19370 entries, 0 to 19369
Data columns (total 12 columns):
 #   Column               Non-Null Count  Dtype
---  ------               --------------  -----
 0   user_id              19370 non-null  object
 1   original_image       19370 non-null  object
 2   face_id              19370 non-null  int64
 3   age                  19370 non-null  object
 4   gender               18591 non-null  object
 5   x                    19370 non-null  int64
 6   y                    19370 non-null  int64
 7   dx                   19370 non-null  int64
 8   dy                   19370 non-null  int64
 9   tilt_ang             19370 non-null  int64
 10  fiducial_yaw_angle   19370 non-null  int64
 11  fiducial_score       19370 non-null  int64
dtypes: int64(8), object(4)
memory usage: 1.8+ MB
```

Figure 14.10 Data info.

Print the best five records from the dataframeusing.head() work, confirming the information structure.

total_data.head()

	user_id	original_image	face_id	age	gender	x	y	dx	dy	tilt_ang	fiducial_yaw_angle	fiducial_score
0	30601258@N03	10399646885_67c7d20df9_o.jpg	1	(25, 32)	f	0	414	1086	1383	-115	30	17
1	30601258@N03	10424815813_e94629b1ec_o.jpg	2	(25, 32)	m	301	105	640	641	0	0	94
2	30601258@N03	10437979845_5985be4b26_o.jpg	1	(25, 32)	f	2395	876	771	771	175	-30	74
3	30601258@N03	10437979845_5985be4b26_o.jpg	3	(25, 32)	m	752	1255	484	485	180	0	47
4	30601258@N03	11816644924_075c3d8d59_o.jpg	2	(25, 32)	m	175	80	769	768	-75	0	34

Figure 14.11 First five records.

Plot a visual chart for orientation esteems. This will imagine the difference in information just as outline of what the orientation information holds.

gender['f',m,u']
plt.bar(gender, total_data.gender.value_counts(), align="center", alpha=0.5)
plt.show()

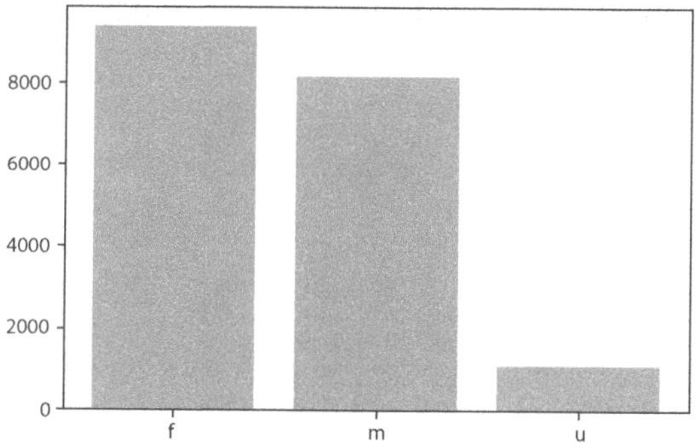

Figure 14.12 Bar chart.

14.7.1.4 Import the Essential Modules for Model Structure

import tensorflow as tf
from keras.models import Sequential
from keras.layers import Conv2D, MaxPooling2D, Activation, Dropout, Flatten,

```
Dense, Dropout, Layer Normalization
from keras.preprocessing.image import ImageDataGenerator, img_to_
array, load_img

path = "AdienceBenchmarkGender AndAgeClassification/faces/"+to-
tal_data.user_id .loc[0]+"/coarse_tilt_aligned face."+str(total_data. face_
id.loc[0])+"" +total_data.original_image.loc[0]
img= load_img(path)
plt.imshow(img)
plt.show()
```

Figure 14.13 Sample data image.

14.7.1.5 Utilize the Important Information and Guide Them

```
imp_data total_data[['age', 'gender','x','y','dx' 'dy']].copy()
imp_data.info()
img_path = []
for row in total_data.iterrows():
        path = "AdienceBenchmarkGender AndAgeClassification/faces/"
        +row[1].user_id=" /coarse_tilt_aligned_face."+str(row[1].face_id)+"
        "+row[1].original_image
img_path.append(path)
imp_data['img_path'] = img_path
imp_data.head()
```

```
<class 'pandas.core.frame.DataFrame'>
RangeIndex: 19370 entries, 0 to 19369
Data columns (total 6 columns):
 #   Column  Non-Null Count  Dtype
---  ------  --------------  -----
 0   age     19370 non-null  object
 1   gender  18591 non-null  object
 2   x       19370 non-null  int64
 3   y       19370 non-null  int64
 4   dx      19370 non-null  int64
 5   dy      19370 non-null  int64
dtypes: int64(4), object(2)
memory usage: 908.1+ KB
```

Figure 14.14 Data info.

	age	gender	x	y	dx	dy	img_path
0	(25, 32)	f	0	414	1086	1383	AdienceBenchmarkGenderAndAgeClassification/fac...
1	(25, 32)	m	301	105	640	641	AdienceBenchmarkGenderAndAgeClassification/fac...
2	(25, 32)	f	2395	876	771	771	AdienceBenchmarkGenderAndAgeClassification/fac...

Figure 14.15 Data samples.

Make an age planning word reference and guide each age information from dataset to an age bunch.

Drop the records that do not have a right age esteem.

age mapping =[('(0, 2)', '0-2'), (2', 0-2), ('3", "0-2'), (4, 6), 4-6), ('(8, 12), 8-13'), ('13", 5-13), (22, 115-20), ((8, 23), 15-20"), (23, 25-32), ((15, 20), 15-20), ((25, 32), 25-32), ((27, 32)", 25-32), (32, 25-32), (34, 25-32'), ('29, 25-32), ((38, 42)", 38-43), (35, 38-43'), ('36', '38-43'), ('42, 48-53'), ('45, 38-43'), ((38, 43), 38-43'), ('(38, 42), 38-43'), ('(38, 48), 48-53), (46, 48-53), ((48, 53), 48-53'), ('55, 48-53'), ('56, 48-53'), (' (60, 100), 60'), ('57, 60-'), ('58', '60+')]

```
age_mapping_dict = {each[0]: each[1] for each in age mapping}
drop labels = []
for idx, each in enumerate(imp_data.age):
        if each == 'None':
                drop_labels.append(idx)
        else:
```

```
        imp_data.age.loc[idx] - age_mapping_dict[each]
imp_data = imp_data.drop(labels-drop_labels, axis-0) #draped None
Values
imp_data.age.value_counts(dropna=False)
```

```
/usr/local/lib/python3.7/dist-packages/pandas/core/indexing.py:670: SettingWithCopyWarni
A value is trying to be set on a copy of a slice from a DataFrame

See the caveats in the documentation: https://pandas.pydata.org/pandas-docs/stable/user_
  iloc._setitem_with_indexer(indexer, value)
25-32     5296
38-43     2776
0-2       2509
8-13      2292
4-6       2140
15-20     1792
48-53      916
60+        901
Name: age, dtype: int64
```

Figure 14.16 Summary of different age groups.

Drop the records that do not have a realized orientation related with them and print the details for the leftover information. The records that have obscure orientation will be dropped in light of the fact that we cannot utilize them to prepare our model.

```
imp_data = imp_data.dropna()
clean_data = imp_data[imp_data.gender = 'u'].copy()
clean_data.info()
```

```
<class 'pandas.core.frame.DataFrame'>
Int64Index: 17452 entries, 0 to 19345
Data columns (total 7 columns):
 #   Column    Non-Null Count  Dtype
---  ------    --------------  -----
 0   age       17452 non-null  object
 1   gender    17452 non-null  object
 2   x         17452 non-null  int64
 3   y         17452 non-null  int64
 4   dx        17452 non-null  int64
 5   dy        17452 non-null  int64
 6   img_path  17452 non-null  object
dtypes: int64(4), object(3)
memory usage: 1.1+ MB
```

Figure 14.17 Data info.

Map the orientation to class names 0 and 1 and print the initial five records to actually look at the trustworthiness of the information.

```
gender_to_label_map={
        'f': 0,
        'm': 1
}
clean_data["gender"] = clean_data["gender"].apply(lambda g:gender_to_
label_map[g])
clean_data.head()
```

	age	gender	x	y	dx	dy	img_path
0	25-32	0	0	414	1086	1383	AdienceBenchmarkGenderAndAgeClassification/fac...
1	25-32	1	301	105	640	641	AdienceBenchmarkGenderAndAgeClassification/fac...
2	25-32	0	2395	876	771	771	AdienceBenchmarkGenderAndAgeClassification/fac...

Figure 14.18 Data after cleaning.

Map the ages to class names 0 to 7 for each age gathering and print the initial five records to check.

```
age_to_label_map={
'0-2' :0,
'4-6':1,
'8-13':2,
'15-20':3,
'25-32':4,
'38-43':5,
'48-53':6,
'60+':7 }
Clean_data['age']=clean_data['age'].apply(lambda age: age_to_label_
map[age])
Clean.data.head()
```

	age	gender	x	y	dx	dy	img_path
0	4	0	0	414	1086	1383	AdienceBenchmarkGenderAndAgeClassification/fac...
1	4	1	301	105	640	641	AdienceBenchmarkGenderAndAgeClassification/fac...
2	4	0	2395	876	771	771	AdienceBenchmarkGenderAndAgeClassification/fac...
3	4	1	752	1255	484	485	AdienceBenchmarkGenderAndAgeClassification/fac...
4	4	1	175	80	769	768	AdienceBenchmarkGenderAndAgeClassification/fac...

Figure 14.19 Data after cleaning.

Make preparing and test datasets with sklearn train test split strategy

For orientation model, resize the pictures to 227,227 proselyte them to numpy exhibit and add them to the test and train dataset. This interaction might occupy a brief period in light of the fact that the dataset is not tiny.

```
x = clean_data[["ing_path']]
y = clean_data[['gender']]
from sklearn.model selection import train_test_split
x_train, x_test, y_train, y_test train_test_split(x, y, test_size=0/3,
random_state=42)

print("Train data shape (3'.format(X_train.shape))
print("Test data shape ().format(X_test.shape))
train images = []
test images = []

for row in X train.iterrows():
        image Image.open(row[1].ing path)
        image image.resize((227, 227))

        data = np.asarray(image)
        train_images.append(data)

for row in X_test.iterrows():

        image Image.open(row[1].ing_path)
        image image.resize((227, 227))
        data= np.asarray(image)
        test images.append(data)
```

```
train images = np.asarray(train_images)
test images = np.asarray(test_images)
print("Train images shape ()".format(train_images.shape))
print("Test images shape {}".format(test images.shape))
```

14.7.1.6 Orientation Model

```
model = Sequential()
model.add(Conv2D(input_shape=(227, 227, 3), filters-95, kernel_size=(7, 7),
   strides=4, padding-"valid", activation="relu")) 3 model.add(Max-
   Pooling2D(pool size (2,2), strides-(2,2)))
model.add(Layer Normalization())
model.add(Conv2D(filters 256, kernel_size (5, 5), strides 1, pad-
ding="same",activation-='relu')) model.add(MaxPooling2D(pool_
size=(2,2), strides (2,2)))
model.add(LayerNormalization())
model.add(Conv2D(filters=256, kernel_size-(3, 3), strides=1, pad-
ding="same activation relu'))
model.add(MaxPooling2D(pool_size=(2,2), strides (2,2)))
model.add(Layer Normalization())
model.add(Flatten()) 13 model.add(Dense(units-512, activation="relu"))
model.add(Dropout(rate=0.25))
model.add(Dense(units 512, activation='relu")) model.
add(Dropout(rate=0.25))
model.add(Dense(units=2, activation="softmax"))
model.summary()
```

```
Model: "sequential"
_________________________________________________________________
Layer (type)                 Output Shape              Param #
=================================================================
conv2d (Conv2D)              (None, 56, 56, 96)        14208
_________________________________________________________________
max_pooling2d (MaxPooling2D) (None, 28, 28, 96)        0
_________________________________________________________________
layer_normalization (LayerNo (None, 28, 28, 96)        192
_________________________________________________________________
conv2d_1 (Conv2D)            (None, 28, 28, 256)       614656
_________________________________________________________________
max_pooling2d_1 (MaxPooling2 (None, 14, 14, 256)       0
_________________________________________________________________
layer_normalization_1 (Layer (None, 14, 14, 256)       512
_________________________________________________________________
conv2d_2 (Conv2D)            (None, 14, 14, 256)       590080
_________________________________________________________________
max_pooling2d_2 (MaxPooling2 (None, 7, 7, 256)         0
_________________________________________________________________
layer_normalization_2 (Layer (None, 7, 7, 256)         512
_________________________________________________________________
flatten (Flatten)            (None, 12544)             0
_________________________________________________________________
dense (Dense)                (None, 512)               6423040
_________________________________________________________________
dropout (Dropout)            (None, 512)               0
_________________________________________________________________
dense_1 (Dense)              (None, 512)               262656
_________________________________________________________________
dropout_1 (Dropout)          (None, 512)               0
_________________________________________________________________
dense_2 (Dense)              (None, 2)                 1026
=================================================================
Total params: 7,906,882
Trainable params: 7,906,882
Non-trainable params: 0
_________________________________________________________________
```

Figure 14.20 Model summary.

14.7.1.7 Train the Model

callback = tf.keras.callbacks.EarlyStopping(monitor="loss', patience=3)
#Callback for earlystopping

model.compile(optimizer adam, loss-tf.keras.losses
.SparseCategoricalCrossentropy(from_logits=True),metrics=['accuracy']) 3 history = model.fit(train_images, y_train, batch_size=32, epochs-25,validation_data=(test images, y_test), callbacks=[callback])

print("++")
model.save("gender_model25.h5")

```
Epoch 1/25
382/382 [==============================] - 11s 27ms/step - loss: 0.6899 - accuracy: 0.5414 - val_loss: 0.6925 - val_accuracy: 0.5241
Epoch 2/25
382/382 [==============================] - 10s 26ms/step - loss: 0.6909 - accuracy: 0.5360 - val_loss: 0.6926 - val_accuracy: 0.5241
Epoch 3/25
382/382 [==============================] - 10s 26ms/step - loss: 0.6910 - accuracy: 0.5345 - val_loss: 0.6922 - val_accuracy: 0.5241
Epoch 4/25
382/382 [==============================] - 10s 26ms/step - loss: 0.6905 - accuracy: 0.5370 - val_loss: 0.6927 - val_accuracy: 0.5241
Epoch 5/25
382/382 [==============================] - 10s 26ms/step - loss: 0.6900 - accuracy: 0.5404 - val_loss: 0.6926 - val_accuracy: 0.5241
Epoch 6/25
382/382 [==============================] - 10s 27ms/step - loss: 0.6895 - accuracy: 0.5419 - val_loss: 0.6922 - val_accuracy: 0.5241
Epoch 7/25
382/382 [==============================] - 10s 27ms/step - loss: 0.6901 - accuracy: 0.5402 - val_loss: 0.6925 - val_accuracy: 0.5241
Epoch 8/25
382/382 [==============================] - 10s 27ms/step - loss: 0.6894 - accuracy: 0.5447 - val_loss: 0.6921 - val_accuracy: 0.5241
Epoch 9/25
382/382 [==============================] - 10s 27ms/step - loss: 0.6899 - accuracy: 0.5426 - val_loss: 0.6922 - val_accuracy: 0.5241
Epoch 10/25
382/382 [==============================] - 10s 27ms/step - loss: 0.6909 - accuracy: 0.5340 - val_loss: 0.6930 - val_accuracy: 0.5241
Epoch 11/25
382/382 [==============================] - 10s 27ms/step - loss: 0.6891 - accuracy: 0.5452 - val_loss: 0.6923 - val_accuracy: 0.5241
++++++++++++++++++++++++++++++++++++++++++++++++++++++++++++++++++++++++++++++++++++++++++++++++++++++++++++++++++++++++
```

Figure 14.21 Results showing loss and accuracy.

test_loss, test_acc = model.evaluate(test images, y_test, verbose=2)
print(test acc)

14.7.1.8 Make Preparing and Testing Split for Age Information

x = clean_data[[| Log_path']]
y = clean_data[[age]]
from sklearn.model_selection import train_test_split
X_train, X_test, y train, y test train_test_split(x, y, test_size=3, random_state=42)

```
print("Train data shape ().format(X_train.shape))
print("Test data shape (format(X_test_shape))

train images = []
test images = []

for row in X train.iterrows():
        image Image.open(row[ 1].img path)
        image image.resize((227, 227)) #Resize the image
        data= np.asarray(image)
        train images.append(data)

for row in X_test.iterrows():
        image = Image.open(row[1].img path)
        image.resize((227, 227)) #Resize the image
        data = np.asarray(image)
        test images.append(data)

train images = np.asarray(train images)
test images = np.asarray(test images)

print("Train images shape ().format(train_images.shape))
print("Test images shape () format(test_images.shape))
```

14.7.1.9 Age Model

```
model Sequential()
model.add(Conv20(input_shape=(227, 227, 3), filters 06, kernel_size=(7,
7), strides=4, padding-valid, activation="relu"))
model.add(MaxPooling20(pool_size=(2,2), strides (2,2)))
model.add(LayerNormalization())

model add(Conv20(filters-250, kernel_size=(3, 5), strides, padding sase,
activation ="relu'))
model.add(MaxPooling2D(pool_size=(2,2), strides (2,2)))
model.add(LayerNormalization())

model add(Conv20(filters 256, kernel_size=(3,), strides, padding="same
activation
="relu'))
model add(MaxPooling2D(pool_size=(-), strides (2.3)))
model.add(Layer Normalization())
```

```
model add(Flatten())

model.add(Dense (units=512, activation=”relu”))
model.add(Dropout(rate=0,25))

model.add(Dense (units=512, activation=relu”))
model.add(Dropout(rate=0.25))
model.add(Dense (units=8, activation “softma”)
model.summary()
callback = tf.keras.callbacks. EarlyStopping (monitor loss, patience 3)
```

```
Model: "sequential_1"
_________________________________________________________________
Layer (type)                 Output Shape              Param #
=================================================================
conv2d_3 (Conv2D)            (None, 56, 56, 96)        14208
_________________________________________________________________
max_pooling2d_3 (MaxPooling2 (None, 28, 28, 96)        0
_________________________________________________________________
layer_normalization_3 (Layer (None, 28, 28, 96)        192
_________________________________________________________________
conv2d_4 (Conv2D)            (None, 28, 28, 256)       614656
_________________________________________________________________
max_pooling2d_4 (MaxPooling2 (None, 14, 14, 256)       0
_________________________________________________________________
layer_normalization_4 (Layer (None, 14, 14, 256)       512
_________________________________________________________________
conv2d_5 (Conv2D)            (None, 14, 14, 256)       590080
_________________________________________________________________
max_pooling2d_5 (MaxPooling2 (None, 7, 7, 256)         0
_________________________________________________________________
layer_normalization_5 (Layer (None, 7, 7, 256)         512
_________________________________________________________________
flatten_1 (Flatten)          (None, 12544)             0
_________________________________________________________________
dense_3 (Dense)              (None, 512)               6423040
_________________________________________________________________
dropout_2 (Dropout)          (None, 512)               0
_________________________________________________________________
dense_4 (Dense)              (None, 512)               262656
_________________________________________________________________
dropout_3 (Dropout)          (None, 512)               0
_________________________________________________________________
dense_5 (Dense)              (None, 8)                 4104
=================================================================
Total params: 7,909,960
Trainable params: 7,909,960
Non-trainable params: 0
_________________________________________________________________
```

Figure 14.22 Results of “sequential_1”.

```
model.compile(optimizer adam, loss-tf.keras. losses
SparseCategoricalCrossentropy(from logits=True), metrics =['accuracy'])
history = model.fit(train images, y train, batch_size=32, epochs-50, valida-
tion data (test images, y test), callbacks [callback])

model.save("age_model50.h5*)
test_loss, test_acc= model.evaluate(test images, y_test, verbose=2)
print(test_acc)
```

14.7.1.10 Induction on the Prepared Model

```
import keras
import json
import sys
import tensorflow as tf
from keras. layers import Input

import numpy as np
import argparse from wide_resnet import WideResNet

from keras.utils.data_utils import get_file
import face_recognition
```

Then, at that point, the model is stacked as the main advance, and its rundown is printed to confirm the right stacking of the model.

```
gender_model = tf.keras.models.load_model('weights.hdf5')
gender_model.summary()
```

Age planning is made to accurately extricate the age from the model.

```
age_map=[['0-2'], ['4-6'], ['8-13'], ['15-20 ], [ 25-32'], ['38-43],[48 -63'],
['60']]
```

To utilize it with a video document, simply pass the general way of that record in the videocapture work.

```
def detect_face(self):
        cap=cv2.VideoCapture(0)
        while True:
                grb, frame=cap.read()
                gray = cv2.cvtColor(frame, cv2.COLOR_BGR2GRAY)
                if not grb:
                        break
```

On the off chance that there are no faces found in the picture, we show the casing all things considered. Sitting tight for esc key press to break the circle.

```
face locations = face_recognition.face locations (frame)
print(face_locations)
if(face locations==[]):
        cv2.imshow("Gender and age, frame)
        if cv2.waitKey(1) == 27:
                break
```

Else, assuming a face is found in the picture, we make a square shape around that region on the edge. We likewise crop the region that contains the face, which we will pass to the model in the wake of arranging it in the subsequent stages.

```
else:
        cv2.rectangle(frame, (face_locations[0][3], face_locations[0][0]),
        (face_locations[0][], face_locations[@][2]), (255, 200, 0), 2)
        img=frame[face_locations[0][0]-25: face_locations[0][2]-25, face_
        locations[0][3]-25; face locations[0][1]+25]

# predict ages and genders of the detected faces
img2= cv2.resize(img, (64, 64))
img2-np.array([img2]).reshape((1, 64,64,3))
results = self.model.predict(img2)

predicted genders = results[0]
gen="F" if predicted genders[0][0] 0.5 else "M"
ages np.arange(0, 101).reshape(101, 1)
predicted_ages = results[1].dot(ages). flatten()

pred=""
pred-str(int(predicted_ages[0]))+" "+str(gen)
print(pred)
cv2.putText(frame, pred, (face_locations[0][3],face_locations[0][0]),cv2.
FONT_HERSHEY_SIMPLEX,0.7, (2, 255, 255), 2)

cv2.imshow("Gender and age', frame)
cap.release()
if cv2.waitKey(1) == 27:
```

cv2.destroyAllWindows()
# When everything is done, release the capture [3]

14.7.1.11 Age and Gender Detection Project Output

Figure 14.23 Output 1.

Figure 14.24 Output 2.

Figure 14.25 Output 3.

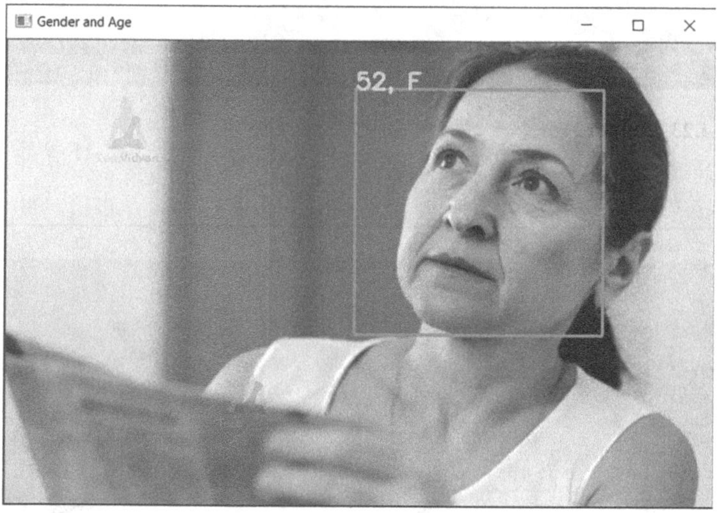

Figure 14.26 Output 4.

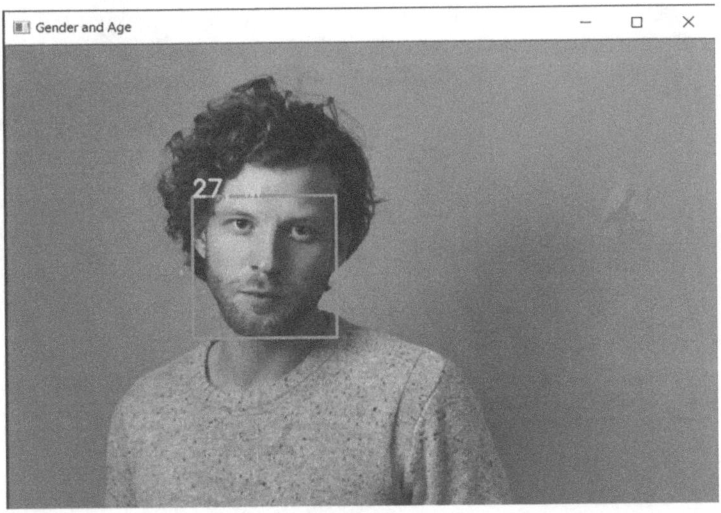

Figure 14.27 Output 5.

References

1. https://medium.com/geekculture/the-challenges-recent-advances-and-future-of-deep-learning-d714304656b
2. https://towardsdatascience.com/5-exciting-deep-learning-advancements-to-keep-your-eye-on-in-2021-6f6a9b6d2406
3. https://techvidvan.com/tutorials/gender-age-detection-ml-keras-opencv-cnn/

15
Advanced Deep Learning Using Julia Programming

15.1 Convolutional Neural Network

```
using Flux
using Flux: @epochs, onehotbatch, onecold, throttle
using Statistics
using Base.Iterators: repeated
# Define your CNN model
function build_model()
   return Chain(
      Conv((3, 3), 1=>16, relu),
      MaxPool((2,2)),
      Conv((3, 3), 16=>32, relu),
      MaxPool((2,2)),
      Conv((3, 3), 32=>64, relu),
      MaxPool((2,2)),
      flatten,
      Dense(64*4*4, 128, relu),
      Dense(128, 10),
      softmax
)
end
# Load your dataset (for example, MNIST)
using Flux.Data.MNIST
imgs = MNIST.images()
labels = MNIST.labels()

# Preprocess your data
X = hcat(float.(reshape.(imgs, :))...) |> gpu
Y = onehotbatch(labels, 0:9) |> gpu
# Split data into training and testing sets
```

Kolla Bhanu Prakash. Machine Learning for Industrial Applications, (277–296) © 2024 Scrivener Publishing LLC

```
train_indices = 1:50000
test_indices = 50001:60000
train_data = (X[:, train_indices], Y[:, train_indices])
test_data = (X[:, test_indices], Y[:, test_indices])

# Build the model
model = build_model() |> gpu

# Define loss function and optimizer
loss(x, y) = Flux.crossentropy(model(x), y)
optimizer = Flux.ADAM(params(model))

# Train the model
epochs = 10
@epochs epochs Flux.train!(loss, train_data, optimizer, cb = throttle(() ->
println("Loss: ", loss(X, Y)), 10))

# Test the model
accuracy(x, y) = mean(onecold(model(x)) .== onecold(y))
acc = accuracy(test_data...)
println("Test accuracy: $acc")
```

15.2 Deep Neural Network Implementation

```
using Flux
using Flux: @epochs, onecold, onehotbatch
using Statistics

# Define your DNN model
function build_model()
  return Chain(
    Dense(784, 128, relu),
    Dense(128, 64, relu),
    Dense(64, 10),
    softmax
)
end

# Load your dataset (for example, MNIST)
using Flux.Data.MNIST
```

```
imgs = MNIST.images()
labels = MNIST.labels()

# Preprocess your data
X = hcat(float.(reshape.(imgs, :))...)
Y = onehotbatch(labels, 0:9)

# Split data into training and testing sets
train_indices = 1:50000
test_indices = 50001:60000

train_data = (X[:, train_indices], Y[:, train_indices])
test_data = (X[:, test_indices], Y[:, test_indices])

# Build the model
model = build_model()

# Define loss function and optimizer
loss(x, y) = Flux.crossentropy(model(x), y)
optimizer = Flux.ADAM(params(model))

# Train the model
epochs = 10
@epochs epochs Flux.train!(loss, train_data, optimizer)

# Test the model
accuracy(x, y) = mean(onecold(model(x)) .== onecold(y))
acc = accuracy(test_data...)
println("Test accuracy: $acc")
```

15.3 Recurrent Neural Network Implementation

```
using Flux
using Flux: onehot, onehotbatch, crossentropy, throttle
using Statistics

# Define your RNN model
function build_model(vocab_size::Int, hidden_size::Int)
    return Chain(
        LSTM(vocab_size, hidden_size),
```

```
        Dense(hidden_size, vocab_size),
        softmax
    )
end

# Define function to generate text data
function generate_text(corpus::String, seq_length::Int)
    data = collect(corpus)
    X = []
    Y = []
    for i in 1:length(data)-seq_length
        push!(X, data[i:i+seq_length-1])
        push!(Y, data[i+seq_length])
    end
    return X, Y
end

# Load and preprocess your dataset (text data)
corpus = "This is an example text used for training RNN model."
vocab = unique(collect(corpus))
vocab_size = length(vocab)
char_to_index = Dict(vocab[i] => i for i in 1:vocab_size)
index_to_char = Dict(i => vocab[i] for i in 1:vocab_size)
seq_length = 10
X, Y = generate_text(corpus, seq_length)
X = hcat(onehotbatch(X, vocab_size)...)
Y = onehotbatch(Y, vocab_size)

# Split data into training and testing sets
train_indices = 1:floor(Int, 0.8 * length(X))
test_indices = length(train_indices)+1:length(X)
train_data = (X[:, train_indices], Y[:, train_indices])
test_data = (X[:, test_indices], Y[:, test_indices])
# Build the model
hidden_size = 128
model = build_model(vocab_size, hidden_size)
# Define loss function and optimizer
loss(x, y) = crossentropy(model(x), y)
optimizer = ADAM(params(model), 0.01)

# Train the model
epochs = 10
```

```
@epochs epochs Flux.train!(loss, train_data, optimizer, cb = throttle(() ->
println("Loss: ", loss(X, Y)), 2))
# Test the model
function generate_text(model, seed::String, length::Int)
   current_char = seed[end]
   result = seed
   for i in 1:length
      input = onehot(vocab_size, char_to_index[current_char])
      output = model(input)
      idx = argmax(output)
      current_char = index_to_char[idx]
      result *= current_char
   end
   return result
end

seed = "This is"
generated_text = generate_text(model, seed, 100)
println("Generated text:")
println(generated_text)
```

15.4 Long Short-Term Memory Implementation

```
using Flux
using Flux: onehot, onehotbatch, crossentropy, throttle
using Statistics
# Define your LSTM model
function build_model(vocab_size::Int, hidden_size::Int)
   return Chain(
      LSTM(vocab_size, hidden_size),
      Dense(hidden_size, vocab_size),
      softmax
)
end
# Define function to generate text data
function generate_text(corpus::String, seq_length::Int)
   data = collect(corpus)
   X = []
   Y = []
   for i in 1:length(data)-seq_length
```

```
        push!(X, data[i:i+seq_length-1])
        push!(Y, data[i+seq_length])
    end
    return X, Y
end

# Load and preprocess your dataset (text data)
corpus = "This is an example text used for training LSTM model."
vocab = unique(collect(corpus))
vocab_size = length(vocab)
char_to_index = Dict(vocab[i] => i for i in 1:vocab_size)
index_to_char = Dict(i => vocab[i] for i in 1:vocab_size)
seq_length = 10
X, Y = generate_text(corpus, seq_length)
X = hcat(onehotbatch(X, vocab_size)...)
Y = onehotbatch(Y, vocab_size)
# Split data into training and testing sets
train_indices = 1:floor(Int, 0.8 * length(X))
test_indices = length(train_indices)+1:length(X)
train_data = (X[:, train_indices], Y[:, train_indices])
test_data = (X[:, test_indices], Y[:, test_indices])

# Build the model
hidden_size = 128
model = build_model(vocab_size, hidden_size)

# Define loss function and optimizer
loss(x, y) = crossentropy(model(x), y)
optimizer = ADAM(params(model), 0.01)

# Train the model
epochs = 10
@epochs epochs Flux.train!(loss, train_data, optimizer, cb = throttle(() ->
println("Loss: ", loss(X, Y)), 2))
# Test the model
function generate_text(model, seed::String, length::Int)
    current_char = seed[end]
    result = seed
    for i in 1:length
        input = onehot(vocab_size, char_to_index[current_char])
        output = model(input)
```

```
        idx = argmax(output)
        current_char = index_to_char[idx]
        result *= current_char
    end
    return result
end
seed = "This is"
generated_text = generate_text(model, seed, 100)
println("Generated text:")
println(generated_text)
```

15.5 Generative Adversarial Network (GAN) Implementation

```
using Random
using Flux
using Flux: @epochs, mse, throttle
using Statistics
# Define your generator model
function build_generator(noise_size::Int, output_size::Int)
    return Chain(
        Dense(noise_size, 128, relu),
        Dense(128, output_size, σ)
    )
end
# Define your discriminator model
function build_discriminator(input_size::Int)
    return Chain(
        Dense(input_size, 128, relu),
        Dense(128, 1, σ)
    )
end
# Generate random noise
function generate_noise(batch_size::Int, noise_size::Int)
    return randn(batch_size, noise_size)
end
# Train the GAN
function train_gan(generator, discriminator, noise_size::Int, batch_
size::Int, data_loader, epochs::Int)
```

```
    loss_fn = mse
    opt_gen = ADAM(params(generator), 0.001)
    opt_disc = ADAM(params(discriminator), 0.001)
    for epoch in 1:epochs
        for real_data_batch in data_loader
            # Train discriminator
            noise = generate_noise(batch_size, noise_size)
            fake_data_batch = generator(noise)
            discriminator_loss = loss_fn(discriminator(real_data_batch),
            ones(batch_size)) + loss_fn(discriminator(fake_data_batch),
            zeros(batch_size))
            Flux.back!(discriminator_loss)
            Flux.update!(opt_disc)
            # Train generator
            noise = generate_noise(batch_size, noise_size)
            fake_data_batch = generator(noise)
            generator_loss = loss_fn(discriminator(fake_data_batch),
            ones(batch_size))
            Flux.back!(generator_loss)
            Flux.update!(opt_gen)
        end
        println("Epoch $epoch completed")
    end
end
# Example usage
noise_size = 100
data_size = 1000
batch_size = 64
epochs = 10
# Generate fake data
fake_data = rand(data_size, noise_size)
# Build models
generator = build_generator(noise_size, data_size)
discriminator = build_discriminator(data_size)
# Train GAN
train_gan(generator, discriminator, noise_size, batch_size, fake_data,
epochs)
```

15.6 Gated Recurrent Unit (GRU) Implementation

```
using Flux
using Flux: onehot, onehotbatch, crossentropy, throttle
using Statistics
# Define your GRU model
function build_model(vocab_size::Int, hidden_size::Int)
   return Chain(
        GRU(vocab_size, hidden_size),
        Dense(hidden_size, vocab_size),
        softmax
   )
end
# Define function to generate text data
function generate_text(corpus::String, seq_length::Int)
   data = collect(corpus)
   X = []
   Y = []
   for i in 1:length(data)-seq_length
        push!(X, data[i:i+seq_length-1])
        push!(Y, data[i+seq_length])
   end
   return X, Y
end
# Load and preprocess your dataset (text data)
corpus = "This is an example text used for training GRU model."
vocab = unique(collect(corpus))
vocab_size = length(vocab)
char_to_index = Dict(vocab[i] => i for i in 1:vocab_size)
index_to_char = Dict(i => vocab[i] for i in 1:vocab_size)
seq_length = 10
X, Y = generate_text(corpus, seq_length)
X = hcat(onehotbatch(X, vocab_size)...)
Y = onehotbatch(Y, vocab_size)
# Split data into training and testing sets
train_indices = 1:floor(Int, 0.8 * length(X))
test_indices = length(train_indices)+1:length(X)
train_data = (X[:, train_indices], Y[:, train_indices])
test_data = (X[:, test_indices], Y[:, test_indices])
# Build the model
```

```
hidden_size = 128
model = build_model(vocab_size, hidden_size)
# Define loss function and optimizer
loss(x, y) = crossentropy(model(x), y)
optimizer = ADAM(params(model), 0.01)
# Train the model
epochs = 10
@epochs epochs Flux.train!(loss, train_data, optimizer, cb = throttle(() ->
println("Loss: ", loss(X, Y)), 2))
# Test the model
function generate_text(model, seed::String, length::Int)
current_char = seed[end]
result = seed
for i in 1:length
        input = onehot(vocab_size, char_to_index[current_char])
        output = model(input)
        idx = argmax(output)
        current_char = index_to_char[idx]
        result *= current_char
   end
   return result
end
seed = "This is"
generated_text = generate_text(model, seed, 100)
println("Generated text:")
println(generated_text)
```

15.7 Radial Basis Function Network (RBFN) Implementation

```
using LinearAlgebra
# Define the Radial Basis Function
function radial_basis_function(x, c, σ)
   return exp(-norm(x - c)^2 / (2 * σ^2))
end
# Define the Radial Basis Function Network (RBFN)
struct RBFN
   centers::Matrix{Float64}
   weights::Vector{Float64}
```

```
    σ::Float64
end
# Constructor function for RBFN
function RBFN(centers::Matrix{Float64}, σ::Float64)
    return RBFN(centers, zeros(size(centers, 2)), σ)
end
# Train the RBFN using least squares method
function train_rbfn!(rbfn::RBFN, X::Matrix{Float64}, Y::Vector{Float64})
    Φ = [radial_basis_function(X[:, i], rbfn.centers[:, j], rbfn.σ) for i in
    1:size(X, 2), j in 1:size(rbfn.centers, 2)]
    rbfn.weights = Φ \ Y
end
# Predict function for RBFN
function predict(rbfn::RBFN, x::Vector{Float64})
    Φ = [radial_basis_function(x, rbfn.centers[:, j], rbfn.σ) for j in
    1:size(rbfn.centers, 2)]
    return sum(rbfn.weights .* Φ)
end
# Example usage
# Generate some sample data
X = randn(2, 100) # Input data
Y = sin.(sum(X, dims=1)) # Output data (sin of sum of input dimensions)
# Define RBFN parameters
num_centers = 10
σ = 0.1
# Initialize RBFN with random centers
centers = randn(2, num_centers)
# Create and train RBFN
rbfn = RBFN(centers, σ)
train_rbfn!(rbfn, X, Y)
# Test RBFN
x_test = [0.5, 0.5]
prediction = predict(rbfn, x_test)
println("Prediction for $x_test: $prediction")
```

15.8 Multilayer Perceptron (MLP) Implementation

```
using Flux
using Flux: onehotbatch, onecold, crossentropy, throttle
using Statistics
```

```
# Define your MLP model
function build_model(input_size::Int, hidden_sizes::Vector{Int}, output_
size::Int)
   layers = []
   push!(layers, Dense(input_size, hidden_sizes[1], relu))
   for i in 2:length(hidden_sizes)
       push!(layers, Dense(hidden_sizes[i-1], hidden_sizes[i], relu))
   end
   push!(layers, Dense(hidden_sizes[end], output_size))
   return Chain(layers...)
end
# Load and preprocess your dataset
# Example: MNIST
using Flux.Data.MNIST
imgs = MNIST.images()
labels = MNIST.labels()
# Preprocess your data
X = hcat(float.(reshape.(imgs, :))...)
Y = onehotbatch(labels, 0:9)
# Split data into training and testing sets
train_indices = 1:50000
test_indices = 50001:60000
train_data = (X[:, train_indices], Y[:, train_indices])
test_data = (X[:, test_indices], Y[:, test_indices])
# Define the model architecture
input_size = size(X, 1)
hidden_sizes = [64, 32] # Adjust the number of neurons in hidden layers
as needed
output_size = size(Y, 1)
# Build the model
model = build_model(input_size, hidden_sizes, output_size)
# Define loss function and optimizer
loss(x, y) = crossentropy(model(x), y)
optimizer = ADAM(params(model), 0.01)
# Train the model
epochs = 10
@epochs epochs Flux.train!(loss, train_data, optimizer, cb = throttle(() ->
println("Loss: ", loss(X, Y)), 5))
# Test the model
accuracy(x, y) = mean(onecold(model(x)) .== onecold(y))
acc = accuracy(test_data...)
println("Test accuracy: $acc")
```

15.9 Self-Organizing Map (SOM) Implementation

```
using Random
# Define a structure to represent the Self-Organizing Map
struct SOM
   grid::Matrix{Vector{Float64}}
   input_size::Int
   grid_size::Tuple{Int, Int}
   learning_rate::Float64
end
# Initialize the SOM with random weights
function initialize_som(input_size::Int, grid_size::Tuple{Int, Int})
   grid = [rand(input_size) for _ in 1:grid_size[1]*grid_size[2]]
   return SOM(grid, input_size, grid_size, 0.1)
end
# Find the best matching unit (BMU) for a given input vector
function find_bmu(som::SOM, input::Vector{Float64})
   bmu_index = argmin([norm(input - som.grid[i]) for i in 1:length(som.
   grid)])
   return bmu_index, som.grid[bmu_index]
end
# Update the weights of the SOM
function update_som!(som::SOM, input::Vector{Float64}, bmu_index-
::Int, iteration::Int, max_iterations::Int)
   radius = max(som.grid_size...) * exp(-iteration / max_iterations)
   for i in 1:length(som.grid)
        distance = abs(i - bmu_index)
        if distance <= radius
        influence = exp(-(distance^2) / (2 * radius^2))
        som.grid[i] += som.learning_rate * influence * (input - som.grid[i])
        end
   end
end
# Train the SOM using the given inputs
function train_som!(som::SOM, inputs::Matrix{Float64}, max_iterations::Int)
   for iteration in 1:max_iterations
      for input in eachcol(inputs)
         bmu_index, bmu = find_bmu(som, input)
```

```
            update_som!(som, input, bmu_index, iteration, max_iterations)
        end
    end
end
# Example usage
# Generate some sample data
input_size = 2
inputs = rand(input_size, 100)
# Define SOM parameters
grid_size = (10, 10)
max_iterations = 100
# Initialize and train the SOM
som = initialize_som(input_size, grid_size)
train_som!(som, inputs, max_iterations)
# Print the learned grid
println("Learned SOM grid:")
for i in 1:length(som.grid)
    println("Unit $i: $(som.grid[i])")
end
```

15.10 Deep Belief Network (DBN) Implementation

```
using Flux
using Random
# Define a structure for Restricted Boltzmann Machines (RBMs)
struct RBM
    W::Matrix{Float64}
    bv::Vector{Float64}
    bh::Vector{Float64}
end
# Constructor function for RBM
function RBM(visible_size::Int, hidden_size::Int)
    W = randn(hidden_size, visible_size)
    bv = zeros(visible_size)
    bh = zeros(hidden_size)
    return RBM(W, bv, bh)
end
# Gibbs sampling function for RBM
function gibbs_sample(rbm::RBM, v::Vector{Float64}, k::Int)
    v_new = copy(v)
```

```
    h_new = bernoulli.(sigmoid.(rbm.W * v_new .+ rbm.bh))
    for i in 1:k
        v_new = bernoulli.(sigmoid.(rbm.W' * h_new .+ rbm.bv))
        h_new = bernoulli.(sigmoid.(rbm.W * v_new .+ rbm.bh))
    end
    return v_new
end
# Train the RBM using Contrastive Divergence (CD-k)
function train_rbm!(rbm::RBM, data::Matrix{Float64}, learning_rate::-
Float64, k::Int, epochs::Int)
    num_examples = size(data, 2)
    for epoch in 1:epochs
        for i in 1:num_examples
            v0 = data[:, i]
            vk = gibbs_sample(rbm, v0, k)
            # Compute gradients
            dW = sigmoid.(rbm.W * v0) * transpose(v0) - sigmoid.(rbm.W *
            vk) * transpose(vk)
            rbm.W .+= learning_rate * dW
            rbm.bv .+= learning_rate * (v0 - vk)
            rbm.bh .+= learning_rate * (sigmoid.(rbm.W * v0) - sigmoid.(rbm.
            W * vk))
        end
    end
end
# Define a Deep Belief Network (DBN) as a stack of RBMs
struct DBN
    rbms::Vector{RBM}
end
# Constructor function for DBN
function DBN(hidden_layer_sizes::Vector{Int})
    rbms = []
    visible_size = 0
    for hidden_size in hidden_layer_sizes
    rbm = RBM(visible_size, hidden_size)
    push!(rbms, rbm)
    visible_size = hidden_size
end
return DBN(rbms)
end
# Fine-tuning function for DBN using backpropagation
```

```
function fine_tune_dbn!(dbn::DBN, data::Matrix{Float64}, labels::Matrix-
{Float64}, learning_rate::Float64, epochs::Int)
    num_layers = length(dbn.rbms)
    input_data = data
    for i in 1:num_layers
        rbm = dbn.rbms[i]
        train_rbm!(rbm, input_data, learning_rate, 1, epochs)
        input_data = sigmoid.(rbm.W * input_data .+ rbm.bh)
    end
    # Define and train a final layer (e.g., softmax layer) using backpropagation
    # Example:
    # model = Chain(Dense(input_size, output_size, sigmoid), Softmax)
    # loss(x, y) = crossentropy(model(x), y)
    # opt = ADAM(params(model), learning_rate)
    # Flux.train!(loss, [(data, labels)], opt, epochs=epochs)
end
# Example usage
# Generate some sample data
data_size = 1000
input_size = 10
hidden_layer_sizes = [20, 30]
output_size = 2
data = rand(input_size, data_size)
labels = rand(output_size, data_size)
# Initialize and train the DBN
dbn = DBN(hidden_layer_sizes)
fine_tune_dbn!(dbn, data, labels, 0.01, 10)
```

15.11 Restricted Boltzmann Machine (RBM) Implementation

```
using Random
# Define a structure for Restricted Boltzmann Machines (RBMs)
struct RBM
    W::Matrix{Float64}
    bv::Vector{Float64}
    bh::Vector{Float64}
end
# Constructor function for RBM
```

```
function RBM(visible_size::Int, hidden_size::Int)
   W = randn(hidden_size, visible_size)
   bv = zeros(visible_size)
   bh = zeros(hidden_size)
   return RBM(W, bv, bh)
end
# Sigmoid function
σ(x) = 1.0 / (1.0 + exp(-x))
# Sample from a Bernoulli distribution
bernoulli(p) = rand() < p ? 1.0 : 0.0
# Gibbs sampling function for RBM
function gibbs_sample(rbm::RBM, v::Vector{Float64}, k::Int)
   v_new = copy(v)
   h_new = [bernoulli(σ(rbm.W[i, :] · v_new + rbm.bh[i])) for i in
   1:length(rbm.bh)]
   for i in 1:k
      v_new = [bernoulli(σ(rbm.W[:, j] · h_new + rbm.bv[j])) for j in
      1:length(rbm.bv)]
      h_new = [bernoulli(σ(rbm.W[i, :] · v_new + rbm.bh[i])) for i in
      1:length(rbm.bh)]
   end
   return v_new, h_new
end
# Train the RBM using Contrastive Divergence (CD-k)
function train_rbm!(rbm::RBM, data::Matrix{Float64}, learning_rate::-
Float64, k::Int, epochs::Int)
   num_examples = size(data, 2)
   for epoch in 1:epochs
      for i in 1:num_examples
         v0 = data[:, i]
         v_k, h_k = gibbs_sample(rbm, v0, k)
         # Compute gradients
         dW = σ.(rbm.W * v0) * transpose(v0) - σ.(rbm.W * v_k) *
         transpose(v_k)
         rbm.W .+= learning_rate * dW
         rbm.bv .+= learning_rate * (v0 - v_k)
         rbm.bh .+= learning_rate * (σ.(rbm.W * v0) - σ.(rbm.W * v_k))
      end
   end
end
# Example usage
```

```
# Generate some sample data
data_size = 1000
visible_size = 10
hidden_size = 5
data = rand(0:1, visible_size, data_size)
# Define RBM parameters
learning_rate = 0.1
k = 1
epochs = 10
# Initialize and train the RBM
rbm = RBM(visible_size, hidden_size)
train_rbm!(rbm, data, learning_rate, k, epochs)
```

15.12 Autoencoder Implementation

```
using Flux
using Flux: mse, throttle
using Random
# Define the Autoencoder model
struct Autoencoder
  encoder::Chain
  decoder::Chain
end
# Constructor function for Autoencoder
function Autoencoder(input_size::Int, latent_size::Int)
  encoder = Chain(
    Dense(input_size, 64, relu),
    Dense(64, latent_size, relu)
  )
  decoder = Chain(
    Dense(latent_size, 64, relu),
    Dense(64, input_size, identity)
  )
  return Autoencoder(encoder, decoder)
end
# Define a function for forward pass through the Autoencoder
function (ae::Autoencoder)(x)
  return ae.decoder(ae.encoder(x))
end
# Generate some sample data
```

```
data_size = 1000
input_size = 10
data = randn(data_size, input_size)
# Define Autoencoder parameters
latent_size = 5
learning_rate = 0.01
epochs = 10
# Initialize the Autoencoder
autoencoder = Autoencoder(input_size, latent_size)
# Define loss function and optimizer
loss(x) = mse(autoencoder(x), x)
optimizer = ADAM(params(autoencoder), learning_rate)
# Train the Autoencoder
@epochs epochs Flux.train!(loss, Iterators.repeated((data,), epochs), opti-
mizer, cb = throttle(() -> println("Loss: ", loss(data)), 2))
```

16

Machine Learning for Industrial Applications

16.1 Python Code for Generating Music Using Neural Networks

```
import numpy as np
import pandas as pd
from tensorflow.keras.models import Sequential
from tensorflow.keras.layers import LSTM, Dense, Dropout
from tensorflow.keras.callbacks import ModelCheckpoint
from tensorflow.keras.utils import to_categorical

# Load and preprocess data
data = pd.read_csv('music_data.csv') # Your music dataset
notes = sorted(set(data['note']))
note_to_int = dict((note, number) for number, note in enumerate(notes))
sequence_length = 100 # Adjust according to your data and model
architecture

network_input = []
network_output = []

for i in range(0, len(data) - sequence_length, 1):
    sequence_in = data['note'][i:i + sequence_length]
    sequence_out = data['note'][i + sequence_length]
    network_input.append([note_to_int[char] for char in sequence_in])
    network_output.append(note_to_int[sequence_out])

n_patterns = len(network_input)
network_input = np.reshape(network_input, (n_patterns, sequence_
length, 1))
```

Kolla Bhanu Prakash. Machine Learning for Industrial Applications, (297–316) © 2024 Scrivener Publishing LLC

```
network_input = network_input / float(len(notes))
network_output = to_categorical(network_output)

# Define LSTM model
model = Sequential()
model.add(LSTM(
    256,
    input_shape=(network_input.shape[1], network_input.shape[2]),
    return_sequences=True
))
model.add(Dropout(0.3))
model.add(LSTM(512, return_sequences=True))
model.add(Dropout(0.3))
model.add(LSTM(256))
model.add(Dense(256))
model.add(Dropout(0.3))
model.add(Dense(len(notes), activation='softmax')))
model.compile(loss='categorical_crossentropy', optimizer='adam')

# Train model
checkpoint = ModelCheckpoint(
    'weights.{epoch:02d}-{loss:.4f}.hdf5',
    monitor='loss',
    save_best_only=True,
    mode='min'
)
callbacks_list = [checkpoint]
model.fit(network_input, network_output, epochs=100, batch_size=64,
callbacks=callbacks_list)

# Generate music
start = np.random.randint(0, len(network_input)-1)
int_to_note = dict((number, note) for number, note in enumerate(notes))

pattern = network_input[start]
prediction_output = []

for note_index in range(500): # Adjust according to how long you want
your music to be
    prediction_input = np.reshape(pattern, (1, len(pattern), 1))
    prediction_input = prediction_input / float(len(notes))
```

```
    prediction = model.predict(prediction_input, verbose=0)

    index = np.argmax(prediction)
    result = int_to_note[index]
    prediction_output.append(result)

    pattern = np.append(pattern, index)
    pattern = pattern[1:len(pattern)]

# Save generated music
with open('generated_music.txt', 'w') as output_file:
    for note in prediction_output:
        output_file.write(note + '\n')
```

16.2 Image Caption Generator

```
import numpy as np
import matplotlib.pyplot as plt
import tensorflow as tf
from tensorflow.keras.preprocessing.sequence import pad_sequences
from tensorflow.keras.applications import InceptionV3
from tensorflow.keras.models import Model
from tensorflow.keras.layers import Dense, LSTM, Embedding, Input

# Load the pre-trained InceptionV3 model trained on ImageNet dataset
image_model = InceptionV3(include_top=False, weights='imagenet')

# Remove the classification layer
new_input = image_model.input
hidden_layer = image_model.layers[-1].output
image_features_extract_model = Model(new_input, hidden_layer)

# Load the pre-trained tokenizer
tokenizer = tf.keras.preprocessing.text.Tokenizer()
tokenizer_json = 'tokenizer.json' # Path to your tokenizer JSON file
with open(tokenizer_json, 'r') as json_file:
    tokenizer_config = json_file.read()
tokenizer = tokenizer_from_json(tokenizer_config)
```

```
# Load the pre-trained LSTM model for caption generation
lstm_model = tf.keras.models.load_model('lstm_model.h5')

# Function to generate captions for given images
def generate_caption(image):
    image = preprocess_image(image)
    feature_vector = image_features_extract_model.predict(image)
    start_token = tokenizer.word_index['<start>']
    end_token = tokenizer.word_index['<end>']
    caption = [start_token]

    for i in range(max_caption_length):
        sequence = pad_sequences([caption], maxlen=max_caption_length)
        y_pred = lstm_model.predict([feature_vector, sequence], verbose=0)
        y_pred = np.argmax(y_pred)
        if y_pred == end_token:
            break
        word = index_word[y_pred]
        caption.append(y_pred)

    generated_caption = ' '.join([index_word[i] for i in caption])
    return generated_caption

# Example usage
image_path = 'example_image.jpg' # Path to your image
image = load_and_preprocess_image(image_path)
caption = generate_caption(image)
print("Generated Caption:", caption)

# Function to preprocess image
def preprocess_image(image_path):
    img = tf.io.read_file(image_path)
    img = tf.image.decode_jpeg(img, channels=3)
    img = tf.image.resize(img, (299, 299))
    img = tf.keras.applications.inception_v3.preprocess_input(img)
    return tf.expand_dims(img, 0)

# Function to plot the image and generated caption
def plot_image_caption(image_path, caption):
    img = plt.imread(image_path)
    plt.imshow(img)
    plt.title(caption)
```

```
    plt.axis('off')
    plt.show()

# Example usage
plot_image_caption(image_path, caption)
```

16.3 Image Classification Using Julia

```
using Flux
using Flux.Data: DataLoader
using Flux: onehotbatch, onecold, crossentropy, throttle
using MLDatasets
using Images

# Load dataset (MNIST as an example)
train_X, train_y = MNIST.traindata(Float32)
test_X, test_y = MNIST.testdata(Float32)

# Reshape images to fit model input size (28x28 grayscale images in this
case)
train_X = reshape(train_X, (28, 28, 1, :))
test_X = reshape(test_X, (28, 28, 1, :))

# Define model
model = Chain(
    Conv((2, 2), 1=>16, relu),
    MaxPool((2, 2)),
    Conv((2, 2), 16=>32, relu),
    MaxPool((2, 2)),
    flatten,
    Dense(5*5*32, 10),
    softmax
)

# Define loss function
loss(x, y) = crossentropy(model(x), y)

# Define accuracy function
accuracy(x, y) = mean(onecold(model(x)) .== onecold(y))
```

```
# Prepare data
train_data = DataLoader(train_X, onehotbatch(train_y, 0:9))
test_data = DataLoader(test_X, onehotbatch(test_y, 0:9))

# Define optimizer
opt = ADAM()

# Train model
epochs = 10
evalcb = throttle(() -> @show(accuracy(test_X, test_y)), 10)
Flux.train!(loss, params(model), train_data, opt, cb = evalcb)

# Evaluate model
acc = accuracy(test_X, test_y)
println("Accuracy: $acc")
```

16.4 BERT Text Classifier on Tensor Processing Unit

```
import os
import numpy as np
import tensorflow as tf
import tensorflow_datasets as tfds
import tensorflow_text as text # Import tensorflow_text
from tensorflow.keras.models import Model
from tensorflow.keras.layers import Input, Dropout, Dense
from tensorflow.keras.optimizers import Adam
from tensorflow.keras.callbacks import EarlyStopping
from transformers import BertTokenizer, TFBertModel

# Set the TPU strategy
resolver = tf.distribute.cluster_resolver.TPUClusterResolver(tpu='grpc://'
+ os.environ['COLAB_TPU_ADDR'])
tf.config.experimental_connect_to_cluster(resolver)
tf.tpu.experimental.initialize_tpu_system(resolver)
strategy = tf.distribute.TPUStrategy(resolver)

# Load BERT tokenizer
tokenizer = BertTokenizer.from_pretrained('bert-base-uncased')
```

```
# Define model architecture
def build_model():
    with strategy.scope():
        input_word_ids = Input(shape=(max_seq_length,), dtype=tf.int32,
        name="input_word_ids")
        input_mask = Input(shape=(max_seq_length,), dtype=tf.int32,
        name="input_mask")
        segment_ids = Input(shape=(max_seq_length,), dtype=tf.int32,
        name="segment_ids")

        bert_model = TFBertModel.from_pretrained('bert-base-uncased')
        bert_outputs = bert_model(input_word_ids, attention_mask=input_
        mask, token_type_ids=segment_ids)
        bert_last_hidden_state = bert_outputs.last_hidden_state

        x = Dropout(0.1)(bert_last_hidden_state)
        output = Dense(num_classes, activation='softmax')(x)
        model = Model(inputs=[input_word_ids, input_mask, segment_ids],
        outputs=output)
        model.compile(optimizer=Adam(learning_rate=2e-5), loss='sparse_
        categorical_crossentropy', metrics=['accuracy'])
        return model

# Load dataset
def load_dataset():
    # Load dataset using TensorFlow Datasets (TFDS)
    dataset, info = tfds.load('imdb_reviews/subwords8k', with_info=True,
    as_supervised=True)
    train_dataset, test_dataset = dataset['train'], dataset['test']
    return train_dataset, test_dataset

# Tokenize dataset
def tokenize_dataset(dataset, max_length):
    tokenized_dataset = dataset.map(lambda x_text, x_label: (tokenizer.
    encode(x_text.numpy(), max_length=max_length, truncation=True),
    x_label))
    return tokenized_dataset

# Define parameters
max_seq_length = 128
batch_size_per_replica = 64
```

```
global_batch_size = batch_size_per_replica * strategy.num_replicas_
in_sync
num_classes = 2

# Load and tokenize dataset
train_dataset, test_dataset = load_dataset()
train_dataset = tokenize_dataset(train_dataset, max_seq_length)
test_dataset = tokenize_dataset(test_dataset, max_seq_length)

# Train model
with strategy.scope():
    model = build_model()

early_stopping = EarlyStopping(monitor='val_loss', patience=3, restore_best_
weights=True)
model.fit(train_dataset.batch(global_batch_size), epochs=5, validation_
data=test_dataset.batch(global_batch_size), callbacks=[early_stopping])

# Evaluate model
loss, accuracy = model.evaluate(test_dataset.batch(global_batch_size))
print(f"Test loss: {loss}, Test accuracy: {accuracy}")
```

16.5 Python Program for Reinforcement Learning Agent for Atari 2600

```
import numpy as np
import gym
import tensorflow as tf
from tensorflow.keras.models import Sequential
from tensorflow.keras.layers import Conv2D, Flatten, Dense
from tensorflow.keras.optimizers import Adam
from collections import deque
import random

# Define constants
ENV_NAME = 'Breakout-v0'
MEMORY_SIZE = 1000000
BATCH_SIZE = 32
GAMMA = 0.99
LEARNING_RATE = 0.0001
```

```
EPSILON_MAX = 1.0
EPSILON_MIN = 0.01
EPSILON_DECAY = 0.995
NUM_EPISODES = 1000

# Define DQN model
def build_model(input_shape, num_actions):
    model = Sequential()
    model.add(Conv2D(32, (8, 8), strides=(4, 4), activation='relu',
    input_shape=input_shape))
    model.add(Conv2D(64, (4, 4), strides=(2, 2), activation='relu'))
    model.add(Conv2D(64, (3, 3), activation='relu'))
    model.add(Flatten())
    model.add(Dense(512, activation='relu'))
    model.add(Dense(num_actions, activation='linear'))
    model.compile(loss='mse', optimizer=Adam(lr=LEARNING_RATE))
    return model

# Define replay memory
class ReplayMemory:
    def __init__(self, capacity):
        self.memory = deque(maxlen=capacity)

    def add(self, state, action, reward, next_state, done):
        self.memory.append((state, action, reward, next_state, done))

    def sample(self, batch_size):
        return random.sample(self.memory, batch_size)

# Preprocess frames
def preprocess_frame(frame):
    return np.mean(frame[30:195, 5:155], axis=2)

# Main training function
def train_dqn(env_name, num_episodes):
    env = gym.make(env_name)
    state_size = (84, 84, 4)
    num_actions = env.action_space.n
    model = build_model(state_size, num_actions)
    memory = ReplayMemory(MEMORY_SIZE)
```

```
epsilon = EPSILON_MAX
rewards = []

for episode in range(num_episodes):
    state = preprocess_frame(env.reset())
    state = np.stack((state, state, state, state), axis=2)
    state = np.reshape([state], (1, state.shape[0], state.shape[1], state.
    shape[2]))

    total_reward = 0
    done = False
    while not done:
        if np.random.rand() <= epsilon:
            action = env.action_space.sample()
        else:
            q_values = model.predict(state)
            action = np.argmax(q_values[0])

        next_state, reward, done, _ = env.step(action)
        next_state = preprocess_frame(next_state)
        next_state = np.reshape([next_state], (1, next_state.shape[0], next_
        state.shape[1], 1))
        next_state = np.append(next_state, state[:, :, :, :3], axis=3)
        memory.add(state, action, reward, next_state, done)
        state = next_state

        if len(memory.memory) > BATCH_SIZE:
            minibatch = memory.sample(BATCH_SIZE)
            states, targets = [], []
            for s, a, r, s_next, d in minibatch:
                target = model.predict(s)
                target[0][a] = r if d else r + GAMMA * np.max(model.pre-
                dict(s_next)[0])
                states.append(s[0])
                targets.append(target[0])

            model.fit(np.array(states), np.array(targets), epochs=1, verbose=0)

        total_reward += reward
```

```
        rewards.append(total_reward)
        epsilon = max(EPSILON_MIN, epsilon * EPSILON_DECAY)
        print(f"Episode {episode + 1}/{num_episodes}, Total Reward: {total_reward}, Epsilon: {epsilon}")

    env.close()
    return rewards

# Train DQN agent
rewards = train_dqn(ENV_NAME, NUM_EPISODES)
```

16.6 Python Program for Multi-Lingual ASR with Transformers

```
import torch
from transformers import Wav2Vec2ForCTC, Wav2Vec2Tokenizer, Wav2Vec2Processor
from datasets import load_dataset
from transformers import TrainingArguments, Trainer

# Load multi-lingual dataset
dataset = load_dataset("common_voice", "tr", split="train")

# Load pre-trained model and tokenizer
processor = Wav2Vec2Processor.from_pretrained("facebook/wav2vec2-large-960h")
model = Wav2Vec2ForCTC.from_pretrained("facebook/wav2vec2-large-960h")

# Tokenize audio input and apply CTC loss
def prepare_dataset(batch):
    speech_input = processor(batch["path"], sampling_rate=16_000, return_tensors="pt").input_values
    with processor.as_target_processor():
        labels = processor(batch["sentence"], return_tensors="pt").input_ids
    return {"input_values": speech_input, "labels": labels}

dataset = dataset.map(prepare_dataset)
```

```
# Define training arguments
training_args = TrainingArguments(
    output_dir="./wav2vec2-large-960h-ft",
    num_train_epochs=3,
    per_device_train_batch_size=4,
    evaluation_strategy="steps",
    save_steps=400,
    eval_steps=400,
    logging_steps=400,
    learning_rate=3e-4,
    weight_decay=0.005,
)

# Define trainer
trainer = Trainer(
    model=model,
    args=training_args,
    train_dataset=dataset,
    tokenizer=processor,
)

# Fine-tune the model
trainer.train()
```

16.7 Python Program for Reinforcement Learning for Connect X

```
import numpy as np

# Define the board size
ROWS = 6
COLS = 7

# Define the connect length required to win
CONNECT_LENGTH = 4

# Define player symbols
PLAYER_X = 1
PLAYER_O = -1
EMPTY = 0
```

```
# Function to initialize the board
def initialize_board():
    return np.zeros((ROWS, COLS), dtype=int)

# Function to check if a move is valid
def is_valid_move(board, col):
    return board[ROWS-1][col] == EMPTY

# Function to drop a piece into a column
def drop_piece(board, col, player):
    for row in range(ROWS):
        if board[row][col] == EMPTY:
            board[row][col] = player
            break

# Function to check if there is a winning move
def is_winner(board, player):
    # Check horizontal
    for row in range(ROWS):
            for col in range(COLS - CONNECT_LENGTH + 1):
                if np.all(board[row, col:col + CONNECT_LENGTH] == player):
                    return True

    # Check vertical
    for col in range(COLS):
        for row in range(ROWS - CONNECT_LENGTH + 1):
            if np.all(board[row:row + CONNECT_LENGTH, col] == player):
                return True

    # Check diagonal (top-left to bottom-right)
    for row in range(ROWS - CONNECT_LENGTH + 1):
        for col in range(COLS - CONNECT_LENGTH + 1):
            if np.all([board[row + i][col + i] == player for i in range(CONNECT_
            LENGTH)]):
                return True

    # Check diagonal (top-right to bottom-left)
    for row in range(ROWS - CONNECT_LENGTH + 1):
        for col in range(CONNECT_LENGTH - 1, COLS):
            if np.all([board[row + i][col - i] == player for i in range(CONNECT_
            LENGTH)]):
```

```
        return True

return False

# Function to print the board
def print_board(board):
print(np.flip(board, 0))
  # Function to evaluate the board state
def evaluate(board):
  if is_winner(board, PLAYER_X):
    return 100
  elif is_winner(board, PLAYER_O):
    return -100
  else:
    return 0

# Function to perform Minimax with alpha-beta pruning
def minimax(board, depth, alpha, beta, maximizing_player):
  if depth == 0 or is_winner(board, PLAYER_X) or is_winner(board,
  PLAYER_O) or not np.any(board == EMPTY):
    return evaluate(board)

  if maximizing_player:
    max_eval = -np.inf
    for col in range(COLS):
      if is_valid_move(board, col):
        temp_board = board.copy()
        drop_piece(temp_board, col, PLAYER_X)
        eval = minimax(temp_board, depth - 1, alpha, beta, False)
        max_eval = max(max_eval, eval)
        alpha = max(alpha, eval)
        if beta <= alpha:
          break
    return max_eval
  else:
    min_eval = np.inf
    for col in range(COLS):
      if is_valid_move(board, col):
        temp_board = board.copy()
        drop_piece(temp_board, col, PLAYER_O)
        eval = minimax(temp_board, depth - 1, alpha, beta, True)
```

```
            min_eval = min(min_eval, eval)
            beta = min(beta, eval)
            if beta <= alpha:
                break
        return min_eval

# Function to find the best move using Minimax with alpha-beta pruning
def find_best_move(board, depth):
    best_eval = -np.inf
    best_col = None
    alpha = -np.inf
    beta = np.inf
    for col in range(COLS):
        if is_valid_move(board, col):
            temp_board = board.copy()
            drop_piece(temp_board, col, PLAYER_X)
            eval = minimax(temp_board, depth - 1, alpha, beta, False)
            if eval > best_eval:
                best_eval = eval
                best_col = col
    return best_col

# Main function for playing Connect X against AI
def play_connect_x():
    board = initialize_board()
    print_board(board)

while True:
        # Player's turn
        col = int(input("Your turn (0-6): "))
        if is_valid_move(board, col):
            drop_piece(board, col, PLAYER_O)
            print_board(board)
            if is_winner(board, PLAYER_O):
                print("You win!")
                break
            elif not np.any(board == EMPTY):
                print("It's a tie!")
                break
```

```
        # AI's turn
        print("AI's turn:")
        best_col = find_best_move(board, 4) # Adjust the depth for stronger/
        weaker AI
        drop_piece(board, best_col, PLAYER_X)
        print_board(board)
        if is_winner(board, PLAYER_X):
            print("AI wins!")
            break
        elif not np.any(board == EMPTY):
            print("It's a tie!")
            break

if __name__ == "__main__":
play_connect_x()
```

16.8 Stock Market Analysis and Forecasting Using Deep Learning

```
import numpy as np
import pandas as pd
import matplotlib.pyplot as plt
from sklearn.preprocessing import MinMaxScaler
from tensorflow.keras.models import Sequential
from tensorflow.keras.layers import LSTM, Dense, Dropout
from sklearn.metrics import mean_squared_error

# Load stock price data
data = pd.read_csv('stock_prices.csv') # Replace 'stock_prices.csv' with
your dataset

# Preprocess data
scaler = MinMaxScaler(feature_range=(0, 1))
scaled_data = scaler.fit_transform(data['Close'].values.reshape(-1,1))

# Split data into train and test sets
train_size = int(len(scaled_data) * 0.80)
test_size = len(scaled_data) - train_size
train_data, test_data = scaled_data[0:train_size,:], scaled_data[train_
size:len(scaled_data),:]
```

```
# Convert data into sequences
def create_sequences(data, seq_length):
    X, y = [], []
    for i in range(len(data)-seq_length-1):
        X.append(data[i:(i+seq_length), 0])
        y.append(data[i + seq_length, 0])
    return np.array(X), np.array(y)

seq_length = 100 # Adjust according to your data
X_train, y_train = create_sequences(train_data, seq_length)
X_test, y_test = create_sequences(test_data, seq_length)

# Reshape data for LSTM (samples, time steps, features)
X_train = np.reshape(X_train, (X_train.shape[0], X_train.shape[1], 1))
X_test = np.reshape(X_test, (X_test.shape[0], X_test.shape[1], 1))

# Define LSTM model
model = Sequential()
model.add(LSTM(units=50, return_sequences=True, input_shape=(X_
train.shape[1], 1)))
model.add(LSTM(units=50))
model.add(Dense(units=1))

# Compile model
model.compile(optimizer='adam', loss='mean_squared_error')

# Train model
model.fit(X_train, y_train, epochs=10, batch_size=32)

# Make predictions
train_predict = model.predict(X_train)
test_predict = model.predict(X_test)

# Inverse transform predictions
train_predict = scaler.inverse_transform(train_predict)
y_train = scaler.inverse_transform([y_train])
test_predict = scaler.inverse_transform(test_predict)
y_test = scaler.inverse_transform([y_test])

# Calculate root mean squared error
train_rmse = np.sqrt(mean_squared_error(y_train[0], train_predict[:,0]))
```

```
test_rmse = np.sqrt(mean_squared_error(y_test[0], test_predict[:,0]))
print('Train RMSE:', train_rmse)
print('Test RMSE:', test_rmse)

# Plot results
plt.plot(data['Close'], label='Actual Stock Price')
plt.plot(np.concatenate([train_predict, test_predict]), label='Predicted
Stock Price')
plt.legend()
plt.show()
```

16.9 ASL Recognition with Deep Learning

```
import numpy as np
import matplotlib.pyplot as plt
from tensorflow.keras.models import Sequential
from tensorflow.keras.layers import Conv2D, MaxPooling2D, Flatten,
Dense, Dropout
from tensorflow.keras.preprocessing.image import ImageDataGenerator

# Define constants
IMAGE_SIZE = (64, 64)
BATCH_SIZE = 32

# Define data generators for training and validation sets
train_datagen = ImageDataGenerator(
   rescale=1./255,
   shear_range=0.2,
   zoom_range=0.2,
   horizontal_flip=True,
   validation_split=0.2
)

train_generator = train_datagen.flow_from_directory(
   'asl_alphabet_train', # Path to training data
   target_size=IMAGE_SIZE,
   batch_size=BATCH_SIZE,
   class_mode='categorical',
   subset='training'
)
```

```
validation_generator = train_datagen.flow_from_directory(
    'asl_alphabet_train', # Path to training data
    target_size=IMAGE_SIZE,
    batch_size=BATCH_SIZE,
    class_mode='categorical',
    subset='validation'
)

# Define CNN model
model = Sequential([
    Conv2D(32, (3, 3), activation='relu', input_shape=(IMAGE_SIZE[0],
    IMAGE_SIZE[1], 3)),
    MaxPooling2D(pool_size=(2, 2)),
    Conv2D(64, (3, 3), activation='relu'),
    MaxPooling2D(pool_size=(2, 2)),
    Conv2D(128, (3, 3), activation='relu'),
    MaxPooling2D(pool_size=(2, 2)),
    Flatten(),
    Dropout(0.5),
    Dense(512, activation='relu'),
    Dropout(0.5),
    Dense(29, activation='softmax') # 29 classes for each letter in the ASL
    alphabet
])

# Compile the model
model.compile(optimizer='adam', loss='categorical_crossentropy', metrics=
['accuracy'])

# Train the model
history = model.fit(
    train_generator,
    steps_per_epoch=train_generator.samples // BATCH_SIZE,
    epochs=10,
    validation_data=validation_generator,
    validation_steps=validation_generator.samples // BATCH_SIZE
)

# Plot training history
plt.plot(history.history['accuracy'], label='accuracy')
plt.plot(history.history['val_accuracy'], label='val_accuracy')
```

```
plt.xlabel('Epoch')
plt.ylabel('Accuracy')
plt.legend(loc='lower right')
plt.show()
```

16.10 Build Rick Sanchez Bot Using Transformers

```
pip install transformers
from transformers import GPT2LMHeadModel, GPT2Tokenizer
# Load pre-trained GPT-2 model and tokenizer
tokenizer = GPT2Tokenizer.from_pretrained("gpt2")
model = GPT2LMHeadModel.from_pretrained("gpt2")
# Define conversation history
conversation_history = ""
# Function to generate response
def generate_response(input_text):
    global conversation_history
    input_text = conversation_history + input_text
    input_ids = tokenizer.encode(input_text, return_tensors="pt")
    # Generate response
    with torch.no_grad():
        output = model.generate(input_ids, max_length=100, pad_token_
        id=tokenizer.eos_token_id)
    response = tokenizer.decode(output[0], skip_special_tokens=True)
    # Update conversation history
    conversation_history += input_text + response
    return response
# Start conversation with Rick Sanchez Bot
print("Rick Sanchez Bot: Hello Morty, what's on your mind?")
while True:
    user_input = input("You: ")
    if user_input.lower() == "bye":
        print("Rick Sanchez Bot: Goodbye Morty!")
        break
    response = generate_response(user_input)
    print("Rick Sanchez Bot:", response)
```

Index

Printed in the USA/Agawam, MA
September 16, 2026

910661.098